T5-BCC-153

THE RADIO PAPERS:

FROM KRAB TO KCHU

To
AL KRAMER
who told me to raise hell in radio

and to

MICHAEL BADER
who showed me how to do it

OTHER BOOKS BY LORENZO WILSON MILAM

The Myrkin Papers, 1969

Sex and Broadcasting: A Handbook on Building a Radio Station for the Community, 1975

The Petition Against God: The Full Story Behind the Lansman-Milam Petition Rm -2493, by Rev. A.W. Allworthy, 1976

The Cripple Liberation Front Marching Band Blues, 1984

•

THE RADIO PAPERS:

FROM KRAB TO KCHU

Essays on the Art and Practice of Radio Transmission

LORENZO WILSON MILAM

•

MHO & MHO WORKS 1986 SAN DIEGO, CALIFORNIA

THIS BOOK IS PUBLISHED BY MHO & MHO WORKS AND WAS MANUFACTURED IN THE UNITED STATES OF AMERICA.

FOR A COMPLETE CATALOG OF OTHER BOOKS BEING OFFERED BY MHO & MHO WORKS SEND A STAMPED, SELF ADDRESSED ENVELOPE TO:

MHO & MHO WORKS
POST BOX 33135
SAN DIEGO, CALIFORNIA

COMMERCIAL ORDERS MAY BE PLACED WITH BOOKPEOPLE, 2929 FIFTH AVENUE, BERKELEY, CALIFORNIA 94710. BOOKPEOPLE'S TOLL FREE NUMBER IN THE U.S. EXCEPT CALIFORNIA IS (800) 227-1516, OR IN CALIFORNIA, (800) 624-4466. EAST COAST ORDERS MAY BE PLACED WITH INLAND BOOKS, 22 HEMINGWAY, EAST HAVEN, CONNECTICUT 06512.

LIBRARY OF CONGRESS CATALOGING IN PUBLICATION DATA

MILAM, LORENZO W.
THE RADIO PAPERS, FROM KRAB TO KCHU.

1. RADIO BROADCASTING—UNITED STATES—ADDRESSES, ESSAYS, LECTURES. 2. RADIO PROGRAMS—UNITED STATES—ADDRESSES, ESSAYS, LECTURES. 3. RADIO STATIONS—UNITED STATES—ADDRESSES, ESSAYS, LECTURES. I. TITLE.
HE8698.M528 1985 384.54'0973 85-18897

ISBN: 0-917320-18-2 (CLOTH)
ISBN: 0-917320-19-0 (PAPER)

1 2 3 4 5 6 7 8 9 10 J Q K A

CONTENTS

AUTHOR'S INTRODUCTION

MANY of these essays cover the period 1962-1977. They were written for the program guides of five non-commercial, "community" radio stations: KRAB (Seattle), KBOO (Portland), KDNA (St. Louis), KTAO (Los Gatos), and KCHU (Dallas).

They were written with haste and—especially, those from the early years—without the comforting knowledge that this type of radio would grow and spread so that today there are almost two hundred such stations on the air.

Because of the usual deadlines, there are the usual infelicities of grammar and style that I have taken the liberty of purging. There are, as well, some contradictions and repetitions which I have left standing. Who am I, indeed, to second guess myself two decades past all reason?

As I look them over, I am left with a feeling of pleasure that I had the luck to be in on the early days of (what was then) considered to be revolutionary, even dangerous, communication. In December of 1962, when KRAB went on the air, there were only three other "community" stations in the country. All were considered to be radical and inimical to the peace and survival of The State. The Pacifica stations (New York City, Berkeley, and Los Angeles) were under fairly constant attack from the U.S. Senate Internal Security Subcommittee through the aegis of one Jas. Eastland. The Federal Communications Commission harassed the stations unmercifully—to the point of not even granting small permit requests such as changing a transmitter, or modifying an antenna. (This is a well-known form of bureaucratic nibbling which has the effect of demoralizing the victim by *regulating to death*. It was brought to high art in Germany between 1933 and 1945.)

Thus, the FCC was hardly interested in receiving my two applications for listener-supported radio stations in 1959 (one for Seattle, one for Washington, D.C.) They managed to sit on them for three years, forgotten in the back of the files of the Security Division of the Commission, under the care of some pie-brain by the name of John Harrington. It took major and expensive efforts by attorneys, and the usual signing of the usual loyalty oaths to get one of them out of hock. The Washington, D.C. application died aborning.

For all these reasons, the early essays tend to be somewhat defensive. I wanted to prove to Seattle—and incidentally, to our Peeping Tom government—that ours was a single small voice of reason in a broadcast band otherwise garish and ugly with commercialism and rank anti-intellectualism. We forget, so soon, how hideous American broadcasting in the 50's (NPR wasn't invented until 1970). We forget, so soon, how terrified we were, in those far-off days, of sporting any tinge or hint of radicalism. The ghoul of Joseph McCarthy, after all, had scarce been chased out of the charnel-house of American politics.

So, I give these essays to you with all the appropriate caveats. I am proud of them and the stations that were their spawn. I think we did great radio, with great people, and great camaraderie, on a pitiful budget—some 1/100th of what it takes now to run that miserable Corporation for Public Broadcasting, with all their desultory stations, for a single day.

Now that I have fallen so pleasantly into my dotage, I am glad to be done with this kind of radio. Those years probably went on too long. We did what we set out to do in the first place—that is, to moil the waters and speak the truth. Grant us all the chance for such rowdy accomplishments. *Me non profiteor, secutum esse prae me fero. . .*

LWM
Paradise, California
Summer 1985

Master Oryu asked Master Ryukei: "Everyone has got an origin, where is your origin?"
Ryukei answered: "I had eaten some gruel this morning and now I feel hungry again."
"How is my hand when compared with the hand of Buddha?" Oryu asked.
"Playing biwa under the moon," Ryukei answered.
"How are my feet when compared with those of a mule?" he asked.
"A heron standing in the snow, not of the same color." Ryukei answered.

The Sound of One Hand

THE RADIO PAPERS:

FROM KRAB TO KCHU

THE FIRST KRAB TRANSMITTER.

RADIO PASSION

THERE was a time when we wanted (very seriously) to explain the *aesthetic* of radio—to describe what it is that makes people go so foolish and so broke in order to get involved in radio.

It was a few years ago, and KRAB was but a dull glimmer in the back of our minds (all potential, no kinetic) and, in a letter to one of the co-founders, after a long description of the structure of the station, and the approach, we tried to explain what we thought the aesthetic of radio should be:

". . . we might pick a tall building in the city proper and stick an antenna on top; the advantages of this are obvious—no line charges to the transmitter, no separate charges for studios in one place and tower in another, immediate accessibility to transmitter in case of trouble, and that great smelly driving hum of the transmitter, right in view, so that each moment you *know* that your words are flooding the countryside, bouncing off hills and trees, ramming headlong into cows and people (they don't even feel the run of the words), filtering in somewhere, through a wall, or down the chimney, or struggling (last electron flagging) through the iron struts, into that radio, into that tiny coil of heat that drifts off the back, with the lights: white, ostentatious in front; red, deep, mysterious as the sun-in-my-eyes from the tall mossy tubes behind—and the voice comes flagging in, first jumbled in a mass of megacycles too big for the tiny tubes, then filtered—through rheostat and condenser, through resistor and coil, ground the grid, cool the coil, tote that Wheatstone Bridge, until it comes now, sizzling down the plate here, fluttering over the grid (tiny fingers the voice, like rain on water) draining energy from the universe that tumbles on like a dark night all around the one gleaming eye of the radio (and the smile of the dial, all those numbers as upright as teeth), slipping to pushing and pulling moments in the never-still push-pull circuit, sidling up at last the arches and hams of the loudspeaker, bubbling around the fat condenser, with the two tiny hairs weeping out, being drained of all the scurrilous currents and extrusions of kilowatt, megacycle, ohm, converted at last into a single reed, a single thread of power, that bubbles through the wires (heating them instantly, hands and the thigh), to tumble and burn through the coil, sweeping in countless thousand thousand circles (not getting dizzy, not for one instant), a fine wire-my-nerve coil suspended delicately, like a star, between the gaunt fingers of the magnet (soft the currents: woman fingers) and then it comes, the first insensate vibration, the first shaking of the tender cone, then the burst, the bubbling forth, the rage of violent shake and tumble, the earthquake that stretches the coil to the depths, pulls the cone to the heart of this dark hot circle of light and bubbling charge, and then, as suddenly, when the core is stretched to the final break and tear, the final ruin, as quickly and quietly, the release comes, the tender paper cone (undamaged, only slightly trembling in the warm knowledge pushed past it) falters a moment, and then rights itself in silence."

January 1963

THE FIRST PROGRAM GUIDE

MANY persons are receiving our Program Listing for free for we are attempting to convince these individuals to help us live by subscribing to KRAB.

Our program list alone, or a random hearing of a few of our programs, cannot begin to describe our philosophy of broadcasting, our own peculiar "fairness doctrine." For in our music, talk, discussion, and forum programs, we feel that dissent and opinion are the keynotes. That is why we try to involve so many people in KRAB.

A careful count has shown that in the last two-week period, fifty-five persons appeared on KRAB (either live or on tape—and this does not include BBC programs) to present their views on some aesthetic, social, or political subject. This does not include our regular volunteers (about twelve in number) who make these programs possible by announcing or engineering.

Fifty-five persons, with fifty-five different views. We were delighted that there were so many, and wished that there could be hundreds more—for we see radio as a means to the old democratic concept of the right to dissent: the right to argue, and differ, and be heard.

As long as this country has existed, this right has been more or less accepted. The only problem is the difficulty of *circulation* of these different opinions.

Long ago, when we were very young, the easiest way to give circulation to opinions was by means of the printed word. Print up a few hundred broadsides, pass them out or mail them out.

However, in these days of expensive newsprint, and of expensive mailings, and of expensive manpower, the broadside has died. Perhaps it is the hurry of the times, the plethora of advertising, the jaundice of a dying civilization. Dissent is no longer as cheap as it once was.

We live in a new time, with new and more frightening troubles: and differences of opinion must find a new medium. It would seem, and once it did seem, that radio and television should be that new medium. We cannot help but look with nostalgia on the early pronouncements of those in the development of radio during the twenties. Herbert Hoover, the man—as Commerce Secretary—most responsible for the establishment of the old Federal Radio Commission, could speak with genuine enthusiasm of "the voice, being carried through the ether, throughout the land, into every home—carrying the potentialities of instant communication of thought, drama, music." Everyone was excited, and the journals of the time were ecstatic with the hopes of education, enlightenment, and knowledge—being transmitted everywhere.

But the potentialities of radio were never realized, because of the peculiar set-up of radio in this country; those organizations financially able to support radio had found that drama and poetry and political dissent and economic thought and social questions were not appealing; were not "popular." There were and are few leaders in radio and television—leaders in the BBC sense—and dissent and opinion were shunted aside largely through fear.

With the coming of FM radio, with its relative lack of expense, with the opening of an entirely new spectrum, radio had another chance to go beyond simple entertainment. The FCC realized this by setting aside 20% of the FM band for pure non-commercial stations. And a few idealists recast the idea of dissension and discussion: KPFA in Berkeley, WFMT in Chicago, WGBH in Boston, WRVR in New York.

The idealism of the 20's, and the work of these stations, were the inspiration for KRAB. And that fifty-five people can come to our studios, to say fifty-five different things about as many subjects, means that KRAB is beginning to move towards filling the responsibility abdicated by the commercial broadcasters.

And so we exist, and have no sponsors, and will never have any—even if it means ceasing to exist. Would a sponsor tolerate our recent discussion of television and its effect on the child, where one panelist told a television station director "There is too much crime and violence on television?" Would a sponsor even begin to tolerate our discussion on "The Negro and the New South" that took place here last Sunday night? Would a sponsor look twice at our commentary series—where fourteen individuals of fourteen of the most divergent views are given fifteen minutes to a half-hour bi-weekly to expound their views? We doubt it.

So here we are: with what we think are vital programs. And we have no commercials. And we have many bills. We need money to continue—and the only way for us is to appeal to our listeners: to have each individual listener who finds our programming at all interesting or necessary support us directly by means of yearly subscriptions.

January 1963

PROGRAMMED SILENCE

ACTUALLY, one can be too self-conscious about the sound radio broadcasting medium. And perhaps KRAB is just that. But, when we began all this, we had the theory that experimentation with radio (dramatic effects, thematic continuity, unity) had gone out when commercials came in. We resolved to experiment with ideas: to play with the potentialities of broadcasting much as was done in the 20s and early 30s.

Like silence. We found the jamming of commercials to be repulsive: that is, that once on the air, commercials, music, and talk should be run together so that there would be no respite. The theory is that a pause will send the listener to other stations where there is a greater jam-up.

We didn't like it, so from the first day we were on the air, we made a judicious use of silence: if we found ourselves running ahead of time, we would announce a few moments of silence, turn off the microphone, and do a few busy things around the studio. Or, if we were confused—which we often are—we would pause for a minute or so of silence to gather our scattered thoughts. Or, if there were a particularly stunning piece of music, that came to a crashing halt—we would stop for a minute or so to let the effect die away before breaking in with the voice.

Then there are our volunteer announcers. Actually, they are a necessity to us for we cannot afford to hire a professional to introduce our programs. But further, volunteer announcers bring a variety and freshness to KRAB: they might stumble and worry, but they are alive. For we are appalled by the silky voice of commercial radio and the stereotyped voice of the automated stations.

Sometimes our experiments fail. We felt at one time—for instance—that normal conversation should be carried into announcing. Thus if the announcer could not pronounce "Slenczynska" or "Ockeghem," he could feel free to ask anyone else in the room while he was on the air. It didn't work, and we do it rarely now. The microphone, with its monophonic orientation, made KRAB sound like a hang-out of *radio juvenalia*—which, of course, it well may be.

Another experiment, which was more successful, was the use of the live break. In the midst of a panel discussion show—particularly one that is heated—we often take a five or ten-minute break in the middle so our participants can rest, so our audience can rest, and so that we can open the doors and clear out the smoke (since our one studio has no ventilation.) We used to turn off the microphone and broadcast silence. But one of our engineers—inadvertently—left the microphone on during such a break (only at KRAB could such an endearing mistake occur) and the resulting babble that went out over the air was, we felt, not bad at all. So we began to do it consciously. For instance, during our forum on "Approaches to Medical Care for the Aged," the discussion started off stiffly, courteously. But, at the break, we opened the doors and left the microphone on (warning all in the studio). Some visitors drifted into the studio, they began to ask the participants some questions on what they had said, and the stony courtesy of the first half of the program went out the window. The program grew from that point on—it became an organic thing, a live discussion that would take place in a bar or a home—and only rarely in a radio studio. It was an attempt to erase the frightening "third person" presence of the microphone—to carry sound radio as close to life as possible. And it worked.

So we will continue to experiment in this way: some experiments will be embarrassing and we will drop them as quickly as possible; others, we hope, will be as rewarding as this last one. We will depend upon our listeners to tell us of their feelings on such experiments with the limitless potentialities of the broadcasting medium.

March 1963

GAPS

LAST week, at the end of our third month of broadcasting, we essayed to review the factors that had inspired the establishment of KRAB. In a half hour program, we outlined the history of listener-supported radio in this country—then discussed specifically our own operation.

Altogether, we touched on many facets of our operation: the lack of commercials, the attempt to open a radio frequency to the expression of any and all opinion, the use of volunteer labor, the informality, the lack of money, and our concept of musical programming. In future program listings, we would like to touch on all these topics again and—at this time—the last one.

Our approach to music is the same as our approach to the spoken word programs: that KRAB must fill a *supplementary* role to the other stations broadcasting in the Seattle area. Thus we play no rock and roll because it is already available in quantity on other radio stations; we avoid the classical war horses because they too are well-represented.

A few of the other FM stations are programming some

excellent material: KING-FM with its Archive series, KGMJ with their far-reaching concentration on less well-known musical pieces during at least part of the day, KLSN with their sometimes bold 7-11 morning concert, and, on rare days, KISW in their early evening chamber music concert. Through judicious choosing, one can often find an amazing selection of recorded music, available in few other cities in this country.

However, we feel there are some sad gaps in the music offered—especially in the evening. There is certainly no plethora of Medieval, Renaissance, or Baroque music; there is only just enough chamber music; contemporary music is almost completely lacking—especially of the post-Stravinsky school; although there is some jazz on KING-FM, KLSN, KGMJ (with the excellent Martin Williams show which comes on at an un-excellent hour), KZAM, and KTNT—there is hardly enough to take care of the traditional, pre-modern, and rhythm and blues schools. Ethnic music and folk music of other countries are almost totally ignored, except for one short program on KLSN. And records of any type that are not the highest of the fi—such as 78s or early LPs—are hardly ever played. Not enough highs.

Therefore, we see our function at KRAB as one of filling the gaps—of supplementing the other stations, not competing with them. This means a great deal of pre-Beethoven. This means a greal deal of post-Webern. This means a lot of jazz—either without comment or, in the few jazz-with-comment shows that we do have—a great deal of intelligent comment (and this means more than reading off the record jackets). It also means no small amount of folk and ethnic music of other countries, and at least one program a week drawn from older, strictly lo-fi recordings. In other words, we play the material that would be suicide on the commercial stations but which is sheer delight for us.

March 1963

A YOUNG MAN LEARNING TO BUILD A RADIO, 1922.

SUPER HETRODYNE RECEIVER, 1924.

PANELS

ONE of the nice things about a non-commercial, listener-supported radio station (besides the poverty) is the opportunity to experiment. Because there are no advertisers—which means there is no fear of offending the bread-basket—KRAB is free to dither around, trying out all sorts of new ideas. As was pointed out in an earlier program listing, some of these experiments are profound mistakes, and we abandon them immediately; some are rather successful, so we stick with them.

One of our experiments was in the *form* of forum broadcasting. In the matter of time, for instance, we felt that the major failing of the commercial radio station's panel discussion was that there was always a hurry. "We only have a few minutes left" the moderator says constantly, so the participants feel the rush to get as much said in as little time as possible, and they are often trapped. One finds oneself saying things one does not mean; one feels that one cannot do the best possible job in such a short time. We think of it as the tyranny of the clock.

The forums on KRAB are, we like to think, timeless. We schedule them vaguely for an hour or an hour and fifteen minutes—in the hope that they will run on and on. And forum discussions, we have found, are rather like love: they accordion time, make an hour a minute.

The result of this is a long, and sometimes long-winded, exhaustion of the subject. Individuals participating in these programs, we have found, do not really begin to ignore the presence of the microphone until fifteen or twenty minutes have elapsed. Then, and only then, does a relaxed and truly viable atmosphere enter the discussion. Sometimes this leads to endless repetition of a point; sometimes, it leads to an exciting movement of arguments—as the participants get to know each other and each other's thoughts. Sometimes, interestingly enough, the participants get to know each other and the microphone only too well, and they devolve into rather frenetic and droll interruptions, accusations, and even sulks.

This is what we like. For we see a forum as serving two purposes; first of all, it gives interested parties—interested in a certain subject—a chance to air the question at length: to discuss, disagree, deliberate, and finally arrive at no apparent conclusion. But we also see the forum as giving the listener the opportunity to get to know what sort of people take what sort of sides on any given issue. In a half-hour question period like "Meet the Press," one can only begin to get an inkling of the individuals involved. In our almost two hour discussion, "The Problems of Automation," we felt that out of all the confusion and disagreement, the personalities began to peer through the murk of words.

And is this not part of what radio is all about? In its information programs, for instance, to move into the realm of drama; to make discussion a dramatic crossing of personalities and, in the crossing, to reveal the personal character of those involved. We would be boorish to say that the drama of such programs is the drama of living and being human—but still, in its highly artificial world, radio can produce high drama by having four persons disagree at length over one subject.

April 1963

SELLING KRAB

AS OF this mailing, KRAB has one hundred and fifty-six subscribers—about one for each day of broadcast. These subscribers pay $12 a year to receive our program listing and bear the slings and arrows of our own strange form of experimental broadcasting. Now it goes without saying that we could use several thousand more subscribers. The working conditions here are poor modernization of 19th century sweat-shops. The studio space is so inadequate as to be laughable (for instance, all office work must cease during recording periods: the staff and the telephone have to move into the engineering booth or, on fine days, outside the building). The tape recorders, transmitter, and various pieces of technical equipment break down constantly—not out of any innate desire to be insufferable, but through sheer old age. Our printing machine, operated on a very personal basis (it hides in the bedroom of a houseboat) has, of course, its own personality; late in the day it will turn dull by grinding up reams of very expensive paper into a final dismal bubbling halt.

In moments of extreme good will, we can always ask ourselves if we would have it any other way. And in moments of extreme good will, we can always answer yes. For it would be nice to finish one telephone call without moving a dozen or so times with the telephone. It would be nice to record that great banjo player, or that marvelous poet, or that witty speaker, without having the tape recorder go blooey with tape all over the floor. It'd be sheer heaven to get through one full meal at home without the engineer calling inquiring about "those funny blue lights running around the output tubes." Ancient equipment can be scenic and a damned bore at the same time.

And so we always come to the same question: how do we get more money? And we always come to the same answer: get more subscribers. And we come to the same question: how? And we come to the same answer: "You have to sell yourselves more."

We cannot describe for you the number of people that have come to us with great ideas for SELLING KRAB TO SEATTLE. A respectable board of directors, a respectable program guide, a respectable building, a respectable community-oriented sales pitch to make people realize how great KRAB is. "After all," the bank official confided in us, "in this age one must sell oneself; with modern-day advertising methods, you must attract everyone's attention from a multitude of distractions. Poetry and art and music: you've got to sell 'em like toothpaste. People will eat up culture if they're sold on it."

One hundred fifty-six subscribers and this choice: either sell ourselves or muddle along under the present conditions—perhaps into ultimate abject poverty and the destruction of KRAB. And yet, squeezed into the control room, recording a mad beautiful folk-singer; or avoiding smoke and ozone from a tired-out transmitter; or lying in bed listening to our printer: we think we'd much rather go out of business than be forced to tell any person at any juncture how great KRAB is. We know it's great—even the break-downs and blow-outs are great. But one of the prime displeasures of American civilization is that one must spoil the fun of being grand by putting one's grandness into words—to convince people to buy.

Well, the day that we have to go tell people how great we are, precluding their chance to listen and see for themselves: when it comes to that, we will know that we have failed and the civilization of man has gone mad and wrong and we damn well all might as well be dead.

May 1963

KRAB MYTHS

ONE of the sad things about commercial radio (and educational radio) is the need of the broadcaster to be terribly sober, terribly responsible, terribly important. Advertising accounts and public responsibility only mildly exceed the weight of the stone of Sisyphus and, like Sisyphus, broadcasters refuse to be amused at their own ridiculous behavior and importance. Listeners or potential listeners become a stone to be pushed grimly upwards: if you stop to ponder the humor of it all, you may well be squashed.

We may be realists or we may be martyrs here at KRAB—but we strongly suspect that our listening audience does not number in the millions or hundreds of thousands or thousands or hundreds—and sometimes, on dark wet nights with dark wet programs, we suspect that no one else may be listening besides the engineer, ourselves, and the guy next door who gets us mysteriously on Channel 12. Program participants still insist upon asking whether there was any response to their commentary, or review, or music, or tirade, and we often say "Your wife called up and said it was great," or "Actually, the transmitter was off during the whole program but you seemed so engrossed we didn't want to bother you," or "Haven't you gotten over believing we have listeners, yet?"

Actually we may have a few listeners but we try to ignore them if at all possible. For we find that the thought of broadcasting into the void gives a certain spontaneous quality to our announcing and our actions. KRAB is, we suspect, too young to have its own mythology and yet, even after five months, there is a bit of apocrypha about the station—all of it stemming from the fact of our feeling ourselves free of the millstone of too many listeners.

For instance, is it or is it not true that one evening, when the programs were not too good, and the transmitter kept smoking and we kept popping on and off the air—that the engineer said to hell with it and turned off the transmitter and went off and had a beer and never came back? Or how about the time that one of our commentators got in an argument with a station visitor over Hungary and—the argument going on so long and so well—the engineer simply turned the whole discussion on the air (with certain unfortunate, rowdy results at the end)? Or how about the time our deleted records man wanted to hear "Death Don't Have No Mercy" by Blind Gary Davis at 3 AM and—since we were here typing anyway—he called us from home and we turned on the transmitter and heard "Death Don't Have No Mercy" not once but three times, from 3:15 to 3:30 and then we shut down and went home? Or that ghastly time the power transformer dropped blue sparks all over the building and then collapsed completely and we went back on the air at lower power and did a special remote broadcast of the KRAB staff taking out its aggressions on the self-same, now defunct transformer with hack-saws, hammers, and various blunt instruments? Or—we know this is true, because it happened so recently—last Thursday, when the weather was so fine so suddenly—we took the microphone out on the lawn of the neighbor next door and broadcast some with birds around our hair and ants crawling up our legs.

In any event, this is our way and we are stuck with it until such a day that KRAB becomes rich or famous or gets a million dollars from the Ford Foundation, and we imagine we will continue to avoid being overburdened by listeners as long as possible.

May 1963

TIMOTHY LEARY

ONE of our volunteers happened to mention that he was going over to Central Washington State College to a symposium on youth or education or some such, and mentioned that we might like a tape made. He told us Dr. Tim Leary would be speaking which didn't mean much to us but the thought of getting a tape made on anything to do with youth or education always delights us so we gave the volunteer our Magnecord and five hours of tape and our best wishes that the recorder would work. ("Jiggle this cord, wiggle that light—and if it still doesn't work," our engineer told him, "kick it down there.")

Well, it worked and he brought us a tape of Dr. Leary talking for an hour or so and answering questions for another hour or so. And that tape is what we are all about, we would guess—for in an hour, that tape conveys a whole new world of a man with his thoughts and visions in a whole new dimension.

For maybe Dr. Tim Leary is right or maybe he's wrong or maybe he's crazy, but in an hour he speaks with the confidence of a prophet describing the future. He speaks softly and slowly—one of our listeners said that an hour with his voice "was a constant seduction"—telling his story. He tells us that this is his last contact with American education—that he is leaving the country to go to Mexico to set up a clinic to experiment with the mind and the effect on the mind of the various hallucinatory drugs. "It will be a center for experiment and a center for training—and from this core will spread a web, of first ten, then a hundred, then a thousand persons: all spreading this knowledge of what can be done with the mind, of what the mind is capable." The confidence with which Dr. Leary speaks, the rich clarity with which he sees existing institutions (expressed in almost an off-handed manner), his expression of belief in his vision: all these make this tape, which we obtained almost by accident, an artistic masterpiece.

For, as we have said: maybe he's right or maybe he's wrong or maybe he's crazy: but a man like this with a vision such as he has should be given the chance to be heard by many different persons in many different places. Most of the radio stations we know would avoid Dr. Leary like poison for he speaks a heresy: in an hour he attacks American education, Federal drug laws, college administrators, and—what we must call—the contemporary *static* view of the mind. He advocates a new system of education through drugs, and suggests an end to almost all of man's problems—war, prejudice, poverty, hunger—through intelligent, controlled administration of such drugs.

This is Dr. Leary's view and it should be heard so we are playing it again because we feel it so well fills the role of drama through expression of opinion, and we might find ourselves playing it again, religiously, once a month, for the rest of our natural, unhallucinatory, existence.

June 1963

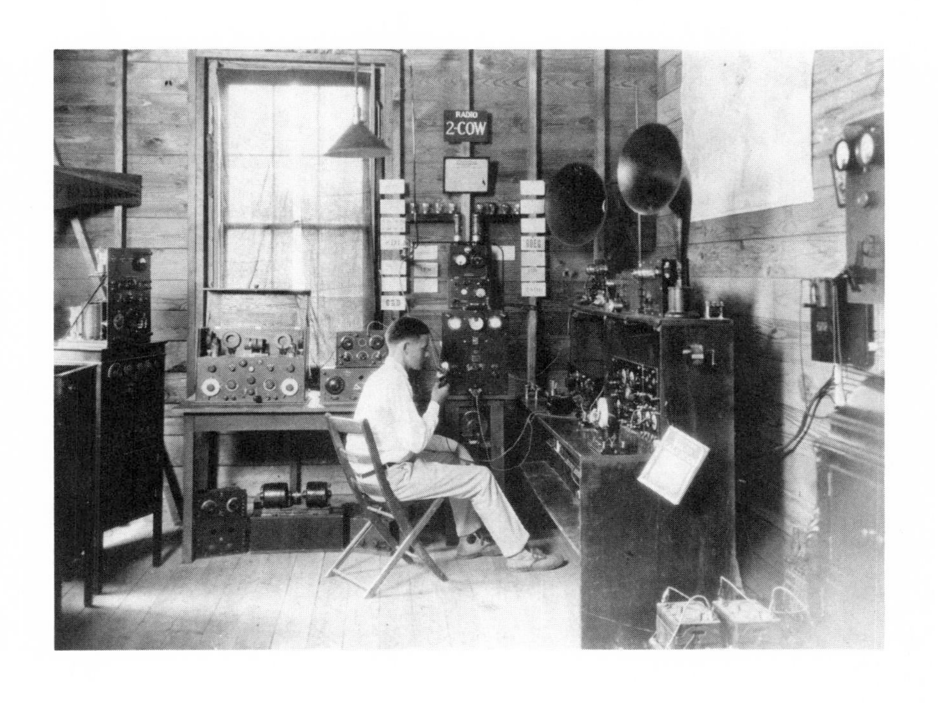

RADIO 2-COW, NEW PLATZ, NEW YORK.

RADIO PARLOR AERIAL, 1922.

SUPPLEMENTARY RADIO

WE HAD the misfortune to tell one of our subscribers and strong supporters that in this program period we were going to schedule two days of Music of India: two days with no relief except for the commentaries. He gave the requisite we-are-doomed look and went on to suggest that during that two day period we could expect five, maybe six listeners: the staff, the music director, the announcer and engineer (captive audience), and the one Indian in town who happened to have an FM radio. "I must say," he said: "this is the ultimate in minority radio." He went on to say that he felt that we played our cards all wrong: we struggle to get the few hundreds or thousands (or is it dozens? or millions?) of faithful listeners, and then we systematically drive them off with such shenanigans as this. "Of course I feel more competitive than you. I feel you should idolize your listeners, not frighten them."

How does a radio station idolize its listeners? Should it accept the rating service, for instance, as the ultimate in democracy—and program what most of the people want most of the time? (Does this then make commercial radio and television the showcase of the democratic system: where profit-taking, majority rule, and pleasure and entertainment march together, hand-in-hand, towards a greater, brighter Tomorrow?) Or even in the more aesthetic, ascetic FM broadcasting, should the musical warhorses and the unctuous announcers convince the listener that he or she is being enculturated without pain, without puzzlement, without distaste?

Most of our subscribers know and accept the principle of non-commercial, listener-supported, minority-orientated, radio. But fewer of them accept the principle of *supplementary* radio: that in every community there should be one outlet that knows and accepts the fact that none of its programs will appeal to all of the people all of the time (or, conversely, that none of its programs will appeal to some of the people much of the time).

Of all our programs, we are most pleased with our commentaries and our music. The commentaries because they give many people a great deal of time and a qualitatively known (although quantitatively unknown) audience who are sure to listen and react. Our music because we know that there are few—if any—radio stations in this country or any other country who would dare to present a two day festival of Medieval Music, or Bach, or Indian Music, or Mozart—or even an hour or two of Chinese Opera, or Electronic Music, or Gagaku, or Recorder Music, or Music of Sunda, or Scot's ballads. Few stations do this because they refuse to appeal to so few people for so long.

We feel that more than a little good comes of all this: it is that we are not here to delight or entertain; but to frighten, entrance, anger, or appall the listener. For coupled with the four or five hundred people that have reacted to our music with disgust, there are the four or five that have said, "Who would ever believe that the music of Sunda could be so great?" or "Music of India—those ragas: they just get inside you, and play all over your nerves." These are the ones who are changed and these are the ones we want to reach.

August 1963

VOLUNTEERS

IT'S a terrible thing to say, but we suppose the truth of the matter is that those who are most influenced by KRAB are not the listeners; rather, it is our small group of volunteer engineers and announcers. After all, they are most often and consistently exposed to the ideas, opinions, and music that flow out over the air because they have to pull the levers and push the buttons that control the flow.

If we had the money, the station's operations could all be run by a paid staff of three: an announcer, an engineer, and a station manager.* This would certainly limit the number of people who run in and out of the building: it would limit the confusion, it would eliminate the considerable number of on-the-air announcing, engineering, and program mistakes. With a set, never-changing staff of two or three, the conditions are fixed; and, as in love, the number of lovers could be set, the conditions of the relationship would be set, the involvement could be fixed.

But KRAB is a prostitute—of sorts: she has many lovers: some faithful, some inconstant, blowing hot-and-cold from day to day. The number of men who come to her is staggering, the number who stay for awhile and leave something of themselves is enough to weaken the purpose of any Moll Flanders. Into this dingy building, during the course of a week, come fifteen or twenty announcers and engineers; then there are the intermittent record cataloguers, typists, painters, sweepers, carpenters; then there are the twenty or so persons who come to tape commentaries, discussions, interviews, readings; and finally there are the twenty or thirty visitors who come to see what she is really like: what this girl they hear so much about really acts, feels, talks, smells like.

But—to shuffle aside this image which is tending towards embarrassment—we like to have this volume of people moving through the station from day to day. Most broadcasters think of it all as one-way communication: you sit there in your dark room, and I'll sit here (with my hot little transmitter), and I will entertain, amuse, perhaps enculturate, maybe even move you. Don't call me, I'll call you.

In spite of this, no one goes into broadcasting unless one is social and likes communication. The excitement of KRAB-type radio is our saying: here is some great stuff—we've put together the best minds we can find to create the best words and music we can find. And if you like it, support us and come help us; if you don't like it, support us and come change us: the door and our minds are always open. And they come: from the University and Seattle and Everett and Chelan and Vancouver and Portland: they come to see her and tell her what's right and what's wrong, and they sometimes work and suffer along with her and try to pay her way and perhaps, if she is lucky enough or if she feels age creeping up on her bones, in time, she *will* become a bit more respectable, a bit less crude, a bit more restrained.

August 1963

**One of our volunteers suggested that the perfect broadcasting medium would be an announcer, an engineer, and a listener: all in the room together—with no one else involved. This puzzling and somewhat Existential view of radio will, no doubt, pop up in one of our future program listings. Meanwhile, we have to go to sleep and think about all the ramifications of it.*

PANEL DISCUSSIONS

WHEN one is setting up discussion programs, one immediately gets to sense the participant's willingness or unwillingness to subject himself to critical questions. Poets, advertising executives, peace workers, doctors, social workers, musicians, internationalists, folk singers, NAACP and CORE members, prohibitionists, birth control proponents: all seem willing to go on the air and publicly defend their views. They seem to feel that there is nothing to hide from, that there is no fact that does not bear scrutiny, no topic that need be hidden from the public.

Panel discussion programs are long and complex affairs to put together: there are calls of initial inquiry, suggestions for participants, calls of confirmation, some calls of cajoling, and the inevitable last-minute substitutions. In all, we have found that it may take as many as forty calls to set up a simple discussion.

The list of panel discussions that never came about—despite all sorts of effort on our part—is long and dismal, and threatens to become longer still. Whatever happened to that discussion on the funeral business: where the big local funeral companies were to appear to discuss some recent charges of over-pricing that have been mentioned in some of the national magazines. We always found the executives "rather busy," or "awaiting the second report of the Washington State Funeral Directors Association," or—later—simply not in. And what about that discussion on recruitment into the Armed Services. Mebbe it was the nervous tone in our voice, or mebbe one should never inquire too deeply into the techniques of recruitment, or mebbe one panel member who is a strong Pacifist does not make for ideal public relations; in any event, the service representatives, at first so eager to go on the air, got more and more nervous and, after a visit to our station, called to say that they were going to be all tied up for some time to come. Why, we ask, should the armed forces, as any other institution in this country, not be willing to subject themselves to critical and pointed questioning?

What happened to all those dentists when we tried to make up a discussion of fluoridation? ("We feel it is better at this time to avoid public debate of the question.") What happened to all the psychologists and pharmacologists when we came to debate the conclusions of Dr. Tim Leary on the question of LSD and the law? What about the local newspaper editors refusing—point-blank—to debate the responsibilities of journalism? ("We feel that we don't have to defend ourselves publicly on this question," said the editor of *The Seattle Times.*) And what did happen to the Air Force representatives when we discussed the book *Fail-Safe*?

We don't ask all these questions out of bitterness but out of wonder. We are constantly (perhaps naïvely) surprised when proponents, opponents, or *status quo* representatives of some controversial issue or group are unwilling to appear before the public and demonstrate the merits and defend the weaknesses of their cause. For we always return to the fact that brought KRAB about: that the public has always shown itself willing to be informed (to some degree); that there is no topic too controversial or disturbing to be aired publicly; that any group should be willing to face strong critical questioning; and that once the information and opinions are presented, the public can and will choose "its own best course."

August 1963

KRAB NIRVANA

ONE of our most loving supporters, a man of no small prestige and accomplishment himself, has a disarming habit of coming into the station and saying "Tell me what you need—*what do you really need now?*"

We find ourselves loath to be boring—to pull out our books (stained with red ink and, no doubt, blood), or to motion dumbly towards the asthmatic, personality-ridden, cantankerous transmitting equipment—and we usually end up asking for some information on possible program material, or perhaps a few names of potential subscribers, or some such. For we know that our never-ending mechanical difficulties can only be solved by a great deal of money, from some extremely generous source and until that person or foundation comes along, we must bank along with what we've got.

In the great pearl-grey vast KRAB-land nirvana of the future, we dream of subscriptions rolling in by the hundreds, people fighting with each other for the chance to be heard over the station, a sleek efficient operation with dozens of eager volunteers; we see visions of clinically new transmitters that transmit, tape recorders that tape, turn-tables that turn, meters that meter, circuits that do not have to be kicked and beaten into surly operation. We see the commentary periods filled with eager fresh vibrant spokesmen for the political left as well as the political right—people dissenting with pleasure, with no question or wonder at the propriety of their right to disagree publicly. We envision a small, subtle yet thoughtful audience, wide-awake, quick to call us on our errors (what few there are) and quicker still to call us to praise our efforts. And we see ourselves, of course, off there in the distance somewhere, on a small bright island in the Mediterranean, sipping Anis, creating massive brilliant programs (on paper: which will be mailed out every week or so), and nodding our heads soporifically in the glory and the wonder of it all.

Once, a very long time ago, probably last week, we were down in Las Vegas—and among all the improbability of the lights and sheer noise of money—we came upon a radio station that went on the air about the same time KRAB did. The owner, in expectation of making piles of money, had spared no cost to put the station on the air. Six new recorders, three turntables, a massive shiny control board, and a brand new transmitter off there somewhere, making quiet coughing sounds as if it really worked. The operator was like the equipment: healthy, well-paid, at ease with himself and all his machinery—sending the commercials and mood-music out over the air as if they were really important.

Sitting there, under all those fluorescent lights, with the wheels ticking mechanically and perfectly, we can't help but admit to a brief pang of envy. No wires all over the floor here! None of the dreadful fear of imminent break-down, the dissolution of everything in a pile of flame and ashes to signify another day of nervous silence as we try to figure what outrageous thing we can do to make everything work for yet another day or so.

It was not a permanent envy: we suppose we're back to normal now, and the basic amusement of our situation has come back to us; still, there was just enough of it so that we thought it would be worth mentioning . . . just this once.

September 1963

QUESTIONNAIRES

ABOUT 25% of the questionnaires we sent out came back and it was fun reading them. The question that evoked most of the wry answers was "What do you find most galling about our operation?" There were a great number complaining about the lateness of programs, the silences, the lateness of arrival of the program guide, the chaotic engineering. One grumbled about "The interminably repeated discussions;" another "the apologetic tone;" another "its existence" (we gather they meant the existence of KRAB's operation at all); others replied "That stuff from Timbuctu, that noise called music," "Your self-satisfied ineptness," "What operation?" "Not finding you when I'm home and you're out," "Lack of supporters," "The telephone," "Tim Leary."

One of the answers that provoked us the most came from one of our Mercer Island subscribers: "Stop fussing vocally about not having advertising on your station: you protest too much. It's all right to stick to your position, but you shouldn't imply that all other stations have sold their souls except yours." We confess this sort of startled us because in a way it echoed a question that has often bothered us: "Why indeed should KRAB rant and rave on the air about being so special when there is often a great deal of excellent material being presented on other FM stations in Seattle?" We've often spoken with envy of the excellent talks and discussions on KUOW and KING-FM; of the music on KISW, KSLN, and especially on KGMJ. Why are we so special that we should demand $12 from our listeners?

Well, there was the usual hum-humming about our commentaries and our live panel discussions—indeed, we know of only three other stations in the country that will permit, much less seek, controversy *without* bias (we've often had to drop into interviews and discussions in a role that is far from what we really believe only to incite the drama of disagreement and strong opinion). There was also the obvious *humanness* that we strive to maintain in our operation: live readings, human mistakes, occasional wit. (As one listener put it, it is the opportunity for the morning announcer to come on the air and say "Good morning . . . grumph" which is, after all, more human than the bright-and-shine artificiality of commercial announcers.)

These are the apparent virtues of KRAB and, we feel, worth $12 if not a million or so. But there was another thought that occurred to us about a week ago. For, very recently, KGMJ-FM was sold for about $45,000. The buyer, also owner of KIXI(AM) and KTVW(TV), might—it is rumored—change the classical programming of KGMJ to a simple repeat of the drivel heard on KIXI; in other words, abandon the excellent classical music concept to (horror of horrors) background-type mood music.

KRAB is non-commercial and always will be—if it can begin to make it financially through the first 18 months; (it never will be anything else under the present ownership set-up). This not only renders it free of advertisers and their appalling restrictions on controversy, it also makes it free of the bids to purchase by commercial interests. Broadcasters are in business to make money. If their programming doesn't pay, they are always willing to sell the business and realize their investment. They could care less about what will happen to the radio station after it is sold.

One simply has to look at the history of AM broadcasting: in the beginning it was an exciting medium, and exciting human things were done with it. People couldn't get over the fact that somebody could talk in one room and be heard hundreds of miles away. Then, shrewd investors realized the enormous profit potential of broadcasting—and that was all she wrote. You can hear the result on any AM radio.

FM is an enormously, *potentially*, profitable medium. Most of the frequencies are gone in the major cities; FM stations are already making scads of money in San Francisco, Los Angeles, Chicago, New York and Boston. All the excitement, all the originality of FM is beginning to disappear in these cities. A Harvard study has predicted that FM set ownership will exceed AM set ownership in 1975.

What we see is that KRAB will be a hedge against this sort of commercial popularity. Local stations will be sold, their programming will move to catering to the tastes of the masses, more money will be invested, more advertisers will be heard on FM, and zap! That's it.

This is why we carry on, talking about the perpetual non-commercial character of this station—for we want our listeners to insure themselves (and ourselves) that there will always be one little nook on the dial for all the rare and unusual and controversial, not just today, or next Sunday, or a year from now, but for as long as the society is free and we are free to dissent in our society.

October 1963

CAR RADIO, 1923.

MINNESOTA, 1942.

MUD AND FLIES

West Seattle
February 7, 1963

Dear KRAB:

I have been listening to your programs for two weeks now and I think your crazy. I go to hear jazz and I get all this African hippo music. I look for some nice background and I get this crazy japanese screaming noise. I want some mellowcrino and end up with buzz-saw grind. You guys are nuts.

And where all the commercials should be there is silence—big holes in the air. Im sure youve gone off the air and what are you doing? Your off the air because your ahead of time. Ive never heard such idiots.

And your commentators if thats what you want to call them. Big Deal. If you dont believe in red blooded americanism you should be shot. I went and fought the second war in the dirt and stink of iwojima and we fought for the flag my buddies getting killed all around and there arms shot off and we fought to make this country what it is today. Free. Up to my neck in mud and stink and I got the disenterry so bad I cant even say now Im well again and eating food fit for pigs living just like pigs. So now Im home and glad of it and we are free from the foreners and then you guys come along. Commentators. They sound like Commies to me and they want to give away all I fought and died for which makes me sick. Where are your guts. If you could be me with the flies over us so thick like we want to choke and the bullets all around like flies and the mud and stink for 3 months so we never saw clean water and hot meat and never knew if we was to see it again either. I fought for my country and didnt get no metal but five losey years in veterans hospital with ciroses of the liver and you guys would give away all my sweat and being away from my girl to some dam commie. You should be shot.

Disgusted.

IT WAS dank and dark back in February and KRAB had been on the air for two months and the machinery had stopped breaking down for awhile and we had a chance to wonder why things were so quiet. Almost as if we weren't here at all. We had put all these programs on the air—the one about broadcasting and the FCC, the interview with the Pacifist who seemed to live in jail, the talk by the reporter who just got back from Cuba, the commentaries from the far Right as well as the far Left, and all the music: Japanese Noh Dramas, African M'Thumpa Drums, Indian Ragas, Lully, Cage, Telemann, Ussachevsky, D'Aquin, Soler, and Blind Lemon Jefferson. All that talk and all that music—and what did we get: some kind letters of praise, a few mild suggestions, and no hate letters, no vilifying calls, no threats even.

Mebbe it was our upbringing or mebbe it was the newspaper stories, but we always thought that controversy on any of the public media brought threats, and rage, and a few rocks, anyway. But not nothing. Nothing at all. "Has the violent fibre of our country totally disappeared?" "Is everyone sitting, bloated and dead, before their television sets?" "Where are the Thomas Paines of yesteryear?"

It was a dank and dark day back in February and KRAB had been on the air for two months and no one seemed to care so we wrote a letter to ourselves, a long dark letter about flies and sweat and dirt and commies and we read it and we liked it pretty good. It seemed to lift the veil a bit (is there really a West Seattle?) and we thought "Man must create his own enemies anyway if he doesn't have any" (does the sad man alone on a rock for years finally people his world with lovers and haters?) and we liked it pretty good so we waited until total darkness when we knew that there was no one out there and we read it with all the misspellings and venom and sat back and waited and when it got more light a letter came in a fine hand from the fellow who was to become our Morning Man: 'Find it hard to believe that "the-stink-and-the-flies-and-the-dead-buddy" fellow was not some sham.'

Which is, we suppose, exactly why he became our morning man; by proving that although hate and anger and violence were quiescent for awhile, cynicism and doubt were still out there and mebbe we had come to a new truth: that the most solid foundation for the New American was neither Revolution nor rage, but a quiet, cunning doubt.

October 1963

SHAMBLES

WE WERE thinking as we were tinkering with the intricacies of this terrifying typewriter, with its wheels that go round and carriage that bounces back by itself and its gears and buttons, as we ruined one master after another through sheer nerves at its massiveness, we were thinking generally of the nature of appearance, the question of the difference it makes in opinion and in real quality. We were content to go on fighting with the vagaries and smudgy letters of our $35 Remington Model One until the other day when the good Dr. Kobler called us and said he heard that we were looking around for something that might give a more workmanlike appearance to our program listing. "I've got an IBM electric here in the office that can make any mistake look as if it really were meant that way," he said.

We admitted a certain distaste for the Remington which made the program sheet look like a journal of pornography put out by some misguided but enthusiastic high-school students. However, we would be far from admitting that we contrived it that way. For, in all our efforts with KRAB, we have admitted a certain distaste for 'appearance.' The station itself, as one of our volunteers suggested, has the general appearance of a disaster area—wires litter the floor, tapes are stacked up to alarming heights, the ceiling periodically falls down, and the fence gate drops off with distressing regularity. If we look quite closely, we will admit that it doesn't have to be this miserable. The ceiling, we guess, with a deal of effort could be glued back together, the tapes could be stacked in a more gentlemanly fashion, and a couple of bolts would take care of that damn gate. But, we wonder, why waste the time? The important thing for any broadcast station is not the beauty of the set up, nor the slickness of the programming—it is the contents of the programming: the thought and work and interest in creating quality, something to stir the mind, something to change just one person, somewhere, once.

The signal of KRAB, our voice over the air, is more often than not a shambles. Engineers are forever leaving the microphone on or putting on the tapes backwards or pushing the wrong buttons. Once again, we would suggest the slickness of the operation is unimportant: we suppose that with a great deal of talking and explaining and nights away from home, we could make the technical aspects of the station almost perfect—but there are so many better things to do; with our understaffing, there are always tapes to be made and calls to put out and hysteria to fight—so we often let the engineers work out the fine points for themselves.

Also, we would admit, somewhat shamefacedly, that we are constantly surprised by the number of people who subscribe to KRAB after they have visited our studios. We used to think that it was our sincere appearance and our apparent aplomb in the face of local ennui, but we've come to think lately that the appearance of the station with its wasted fence and litter is enough to convince the average visitor that perhaps the station does suffer from no small poverty.

Someday, we suppose, if KRAB survives to find subscriptions coming in in droves and foundations fighting with each other to support us, we will have a great cavern of metal and glass with smoothly humming equipment and hundreds of fluorescent lights and acres of gleaming tile, and then we will think of the warmth of our present junk-pile where the important commentary for tonight got lost under the manual for the transmitter (that just burned out) which in turn was left under last week's records which weren't filed because the record cabinet got filled up with all those unedited tapes from Pacifica that got mixed up with the books that we meant to sort out last year. And we will think that the first step towards disaster and conformity came when we abandoned the Remington for all these wheels and keys: but, for the moment, we like the clean letters and the apparent efficiency of it all and are so inspired that we might even get ourselves something to fix our eternally leaky roof.

November 1963

MARATHON

WE suppose that by any contemporary standard the marathon was a rousing success. $1,079 pledged in 42 hours, (over $200 called in in the last hour); about a thousand dollars collected on those pledges.

It was an interesting experience in many ways. The chance to be accessible to the listener every hour of the day and night; the chance to meet many of the people who had heard us for some time and had never been by the station (not a few of them pulled together a few pennies from crocks to give to us); the opportunity to test our own ability to go on and on for hours on end. Further, it gave us the chance to go over our library of tapes and records, to play music we had but never programmed, to hear tapes we had made months ago and had half-forgotten . . . there was that reading, from memory, of Finnegan's Wake, there was the tape of Ian Hamilton Findlay reading his own poetry (he sent it to us from Scotland), there was David Ossman reading from Edmund Wilson, there was the tape of readings of Beun and Celer brought to us by Emile Snyder and for some reason lost until now. There was the chance to experiment . . . the reading of the last letters from Stalingrad (the German soldiers trapped, none were to survive: and these were their last letters), the readings to the music of John Cage. As one listener said, "It's too bad there can't be a Marathon every weekend." As another listener said: "I hear you've made your goal: can I have my $10 back?"

And yet, there was one thing distressing about the whole Marathon. It was the fact that that 42 hours was as close to commercial radio as KRAB has ever come. Every half hour, religiously, sometimes for as long as five minutes, there was a heavy appeal for funds: every half hour, we explained the purpose of the station, the crying need for funds, the necessity for some money in order to survive. Some pleas, expecially in the last hour or so, were so impassioned that they made us want to weep with the pity of us. We learned the degrading lesson of advertising: tell people how good, and noble, and pure you are, for long enough . . . and soon you will come to believe it yourself.

Well, that was yesterday, and today we are less pure, and noble, and good (although slightly more solvent), and we have recognized the hard truth of repetition. And it does not make us too happy. For until the time of the Marathon, we avoided selling ourselves: as we have said so often, KRAB was established to traffic in ideas, not in commerce. We have always limited ourselves to three plugs (or NCSA's, or spots) a day—three quiet explications of the idea of KRAB, and of our need for funds, and the address for the mailing of checks. And, in all that time of moderation, eleven months of quiet appeals, we have only been able to muster slightly over $4,000 in subscriptions—about one a day. In other words, in 42 hours of standard advertising technique, we obtained almost a quarter of what restraint and hesitancy had brought in over the preceding year.

We find it depressing to think what contemporary techniques of advertising have done to Americans; even in our own listeners the advertising klaxons have instilled an automatic blab-off: any appeal for money opens the circuits *unless* it is repeated again, and again, and again. People are dying from an over-profusion of words.

The Marathon has affected us in many ways; for one thing, we never knew our own power: the dry voice of dissent lost somewhere up there on the end of the dial *does* have some listeners, we know that now; further, we know the strength of outrageously repeated appeals; finally, we are tempted to change our whole way of raising money: to go along, quietly, on what we have now—saying nothing about money for a month or two or three; then, when the bank is looking worried and our account is looking vacant, to crash into another Marathon; then, once again, drift along again, until the till runs dry. On and on, drifting between poverty and prosperity, between the non-commercial and the commercial. It would, after all, be a perfect symbol of the schizophrenic nature of listener-supported radio.

November 1963

LISTENING TO THE RADIO IN THE ENDLESS CAVERNS,
NEW MARKET, VIRGINIA.

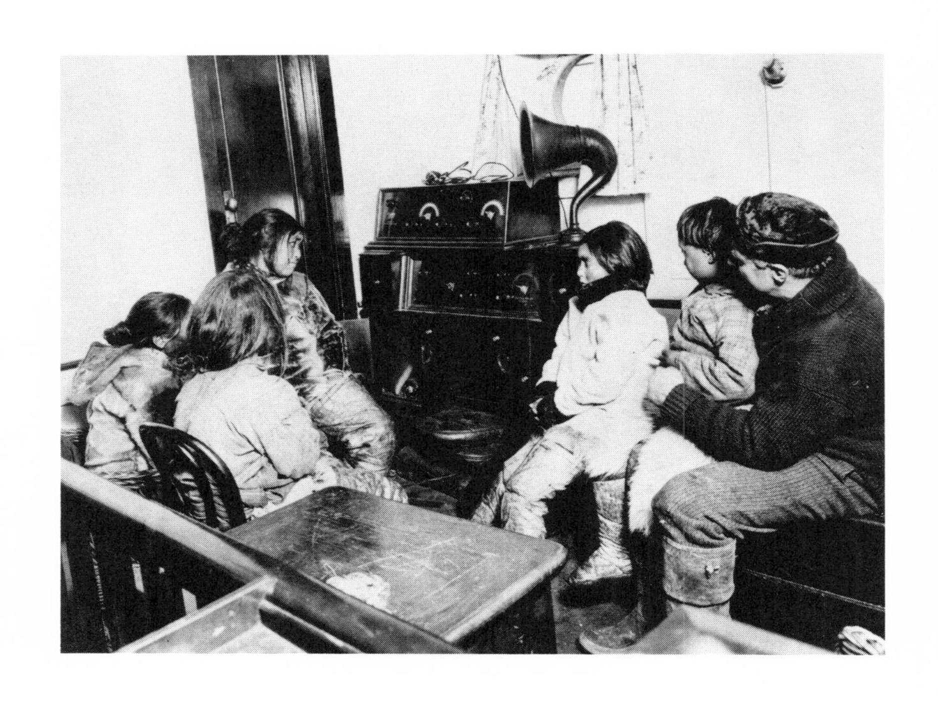

ESKIMOS LISTENING TO A ZENITH, 1926.

THE FIRST YEAR

*"Et semel emissum volat irrevocabile verbum."**

ONE must suppose that people get jaded with any endeavor after awhile: eating rich foods, making love, and running one's own radio station. Diogenes said: "Love is a breadbasket; after awhile, there are only crumbs."

After a year, we would doubt if we are much beyond the crust, and we certainly cannot see the mold. Free-forum, listener-supported radio consumed us then, it consumes us now. We are always staggered with the potentialities of the transmission of sound, and it appears that there are so few groups in this country that have come to appreciate this.

We wish we could remember our first words on the air. We would like to think that it was something significant like "What hath God wrought," or "Veritas cum ego et moritas," but we imagine, more practically, it was a thought shouted darkly, such as "Hey, are we on?" or "I don't hear anything."

Our first real (though short) program was a broadcast sometime late in November, after the equipment had stopped vaporizing our best tubes and hopes; two or three hours of some Indian music, a reading from The New Statesman, and a statement or two about our own ill-defined hopes for this particular station in this particular city. As we best recall, we received three telephone calls that night: one from a gentleman who said our signal was coming in well on 85th Street Northwest, one from a younger sort who said he would appreciate hearing a Fabian record, and one from a lady who asked if we repaired TV sets.

It has been a long year since then. Those who have followed our program guide know the programs we favor the most: the commentaries; the talks of Leary, Graves, Watts and Russell; the music programs with their more than original juxtapositions of music of all times and all places and all cultures; the Saturday morning programs with the readings and the coffee and the music of Sunda and the croissants and the music of Soler. And all the people: Charles, the morning man, who was supposed to go to Africa and ended up in Maryland; Radovitch (who helped us build the station) who crawled to the top of our antenna on sunny days "just to look at the view;" the anonymous man, who was to record a talk for us, but who lost our confidence somehow when he insisted, simply insisted, that he was God; the girl (breathless, all nerves and red cheeks), who interviewed us for a Sociology paper because she felt "we were an important political and social force in Seattle" (dear girl: where are you, where are we, now?); Mike, who had an alarming tendency to go to sleep when making important recordings of dignified people ("someone had to wake me up to tell me to change the tape"); Joe, all of 18 years and already struggling with the world, who heard our presentation of Cage's "Indeterminacy" and so sat on a stump in front of his house for 24 hours ("People would ask me what I was doing there, on that stump, for so long, out in the cold; I would tell them I was protesting, "What are you protesting?' they would ask, 'I dunno'— I would say—'I'm just protesting.' ")

All those people, and all of us, working with the wooden words and the media at hand, trying to communicate. After this year, we say again, as we have said too many times before: It is the desire to communicate, to get some meaning, some words, some thoughts, beyond this room or the next room. We always complain about lack of space, or lack of time, or lack of money—but overall, we never have lost sight of the fact that we are trying to communicate, to tear down the walls somehow, trying to lose sight of the fact that *nil habet infelix paupertas durius in se quam quod ridiculos homines facit.*

December 1963

*Trans. lit.: "Unfortunately for us, when a word is let out of the cage, it cannot be whistled back again nohow."

WARHORSES

CERTAIN of our listeners have intimated that they would not mind having a chance to hear a few of the standard works from time to time—Beethoven's Seventh, Mozart's Thirty-Ninth, Schubert's Unfinished. "There are good war-horses, and there are bad war-horses, and I think you are pretty dumb to ignore them all," said one of our more aggressive volunteers. Another has started bringing in fill material (for the gaps in the programs) like Beethoven Overtures and Chopin Waltzes.

Despite these restless natives, we still are rather militant about the tried and overtried classic. If we were in the radio business to compete with the existing radio stations, then we would lard our programming with all the war-horses we could find. And mood music. And the old and the familiar. For we know it is the familiar and the boorish that commands the largest audience.

Actually, it makes us feel pretty cynical to say that—and we never ceased to be amazed by that: it is simply that most people in this city, on this earth, are really quite dull, are really quite unoriginal, are really quite stuffy. If they were raised on cornbread and pot-likker, then they damn well will eat cornbread and pot-likker until they die or go mad, and no one's going to tell them different. If they heard and liked the 1812 Overture once long ago in a rare mood of enculturation, then they damn well are going to love the 1812 Overture until they die or go mad, and no one's going to tell them differently. To want to hear the strange and original really takes a great deal of energy—to change one's way of eating, or sleeping, to modify one's musical tastes is a lot of work, and few people are going to squeeze out the energy necessary to change.

And sometimes we don't know if they are far from wrong. . . on one of our discussion programs last week, we were discussing the teaching of poetry in the schools with a group of high-school students. They were so militant about the importance of getting poetry across to the average student. "If he doesn't learn to love poetry in high-school he may never learn to love poetry," one of them said. Our question was then: so what; what's so great about learning to love poetry. Poets and writers are so notorious as dissidents, and cynics, and the great unwashed miserable element of civilization: if someone can manage to get through life without reading a bit of poetry, or hearing a note of classical music—then more power to them; at least they will be satisfied with their lot which is more than you can say for the rest of us.

Most people are dull, and unoriginal, and not overly happy, but not unbearably miserable either. Let us hope that every city will create a Warhorse Propagation Society, so that the dull and the unoriginal can be preserved, and the firm foundation of our society may remain. But let us also hope that every city will create an Anti-Warhorse group, which needs so badly to do battle with the dull, and the unoriginal: and thus keep them both alive.

February 1964

TV TURKEY

OUR cantankerous essay of the last program listing was written under the effects of bad Paisano and deadly two-in-the-morning conditions—yet it evoked more comment than some of our properly-reasoned and well-dispatched ideas have stirred over the last year. One listener made vague references to our Fascistic tendencies ("You may think you're better than everyone else, but remember what happened to Hitler," she grumbled). Another wrote a long, intelligent letter, protesting the juxtaposition of the 1812 Overture and Mozart's 39th (not accepting our interpretation of a war-horse as that work which, through the cultural grinding-mill, has been played to death). Someone else suggested that we would not have to worry about creating a Warhorse Propagation Society in Seattle since we already had the Seattle Symphony League. Finally, three of the students who appeared on the discussion of "Poetry in the Schools" suggested that we were pretty dumb since we did not catch the essence of their remarks on Poetry and Truth and that we should not be concerned about whether poetry is Good and Bad for one and etc. etc.

We are often refreshed by what pops out of our typewriter—especially when we are trying to say something different (alert readers can test the relative degree of preparation by comparing the title of the essay with its subject and conclusion); but the concept we managed to phrase last time was pleasing to us: that is, that the great mass of people die before they are twenty and spend their happiest hours delivering tired words to expound tired thoughts; that the greatest joy of most Americans comes when they can take off their ties and scratch their bellies and try their wits against the panel of "What's My Line;" that their greatest conflict is not over whether we should disarm and risk aggression or stay armed and risk accidentally blowing ourselves up—it is whether they should heat up the Frozen Creamed Hamburger Patties in the Electric-Whizz Stove now, or wait and watch "The Beverly Hillbillies" on an empty stomach.

Of course it ain't the fault of the television station owners—the choice as we see it has always been to lead or to follow. If one leads, if one crams onto the television screen even some of the miracles of which the medium is capable, one just don't get so rich so quick. . . was it Cato who said: "To lead is an expensive proposition; almost as expensive as being wise." If one follows, if one crams onto the screen exactly what the rating services say gets the widest audience, one gets an average *18% annual return* on the invested dollar (Newton Minow's figures, not ours). It's all tied up with how quick you want to be rich.

One of our favorite images came in another poorer land, on another poorer continent. We haunted the muddy back streets, and on one—particularly noxious, particularly scenic—we came upon a withered hag in a cane chair with a batch of turkeys. And she was sitting forward, no teeth and all grim determination, cramming some grainy slop down the throat of an unresisting turkey. We asked, in our polite *tourese*, what it was all about, and she craned her head back like a vulture and told us that turkeys 'don't eat too good' and to make them fat for market, one must stuff the craw full three times a day. Certainly a dull way to spend the day, we thought, but the old lady looked as if she could take it, and the turkey, if not exactly thrilled, had that glazed 'stuff-me-if-you-care' appearance about it.

God knows, perhaps our brethren all across this rich land are better off being stuffed in the eye with this cornucopia of TV blah rather than their running out and picketing and starting revolutions or inventing brain gas. But whenever we pass by those multitudes of houses where the curtains are open and the lights are off except for one—that arid, grey-blue light that flickers over the still-warm, humped-over corpses—whenever we see that dark scene, we get that same feeling as when we go past a bar in the morning, or into an all-night movie just before dawn: a feeling of decay and desolation and dying.

February 1964

TO ANNOUNCE

WE MADE up an inter-office memo the other day (our first) for the volunteer announcers. We were pleased at our own business-like approach and so we printed up 50 copies on a borrowed Ditto and tried to hand them out to the announcers so they could know what they were supposed to be doing.

Well, the announcers at KRAB may appear quiet and humble over the air, but a more recalcitrant lot has never been put together under one roof; we guess it could be classified as an Anti-Inter-Office Anything group. We still have 49 copies of our SUGGESTIONS FOR ANNOUNCERS (the other got lost in the trash can). So that the import of our First Inter-Office Memo will not be lost forever, we reprint it for our subscribers in the lucky chance that they too may become volunteer announcers at KRAB and can learn to ignore all directions, posted signs, and printed memos:

"Since what you say over the air represents—for most of the listeners—KRAB, we offer a few suggestions as to your approach to announcing and announcements.

Most of all, remember that you are not trying to sell anything: announcements should be conversational, as if you were speaking to the listener in his or her own livingroom. Projection, excessive seriousness, pompousness, aggressiveness: all are equally as offensive and should be avoided. A bit of wit never hurts, and errors should never be a point of embarrassment: they should be admitted frankly.

The form for all announcements should be consistent and detailed, e.g.:

'That was a reading by the 16th Century English poet Rawdley Bawdley of four of his hexambic poems entitled "The Frog Comes on Little Flat Feet." These poems were written at Brighton Rock during the recent Oceanic Poets Conference sponsored by the S. Brighton Tea and Reading Society. (Pause). This is K-R-A-B, Seattle. (Pause). Next we are going to hear a program of music for the nose-flute, with works of Bawdley, Mawdley, and Sneet. The first work we will hear is the concerto for two nose-flutes, tympani, and organ, by Rawdley Bawdley II. The performers are I Solisti di Sequim, Durdley Flea, conducting.'

If there is some problem with pronunciation, please try to get in touch with someone who should know or a dictionary. If you cannot pronounce a word, don't fake it—spell it out. Never interrupt a concert for anything except music announcements. A concert is to be considered as a continuous whole. You can add a great deal to concerts by 1) Never reading facts about the work or the composer off the record jacket; they tend to be dull, overstated, and often wrong; 2) Doing some research of your own on music or talks to be heard; 3) Reading aloud part of the verse or chorus—or even poetry of the same era or *genre*—if it has intrinsic merit when translated into English.

Commentaries should be introduced by the name of the commentator only: never title the commentary or the commentator—their words should speak for themselves. Refrain from making any judgments—either implied or explicit—about the commentaries. Speakers are on neutral ground here.

Programs may go on late, but never, never early. If you have less than a five-minute gap before the scheduled time of the next program, do a brief reading or have some silence. If we are running more than ten minutes late, mention this fact so that listeners can adjust their schedules accordingly. If a program is unavailable, try to explain why. The listener should be party to some of the problems *we* face.

Plugs (non-commercial spot announcements) are hard to write, we know; but they are absolutely essential to our survival. Always explain WHY someone should support us: it is not enough to say merely that KRAB is non-commercial and starving. Be brief, but not too brief; humble, but not obnoxiously so; and direct. It's a fine line between asking for money and praising the station, but try to avoid the latter. If we are good, the listener will know it . . . if we are not, no amount of saying so will change that fact.

Use pauses judiciously. If you have just finished a stormy piece of music, there is no need to jump right in and announce what it was, even if the engineer is gesticulating insistently. Take your own good time and let the rage die out properly. If you are caught unawares by the end of a piece, pause, collect your thoughts and papers, then announce. Time, telephone, visitors: all these are less important than your appearing calm, wise, and competent over the air."

March 1964

BOY POINTING AT A MICROPHONE, 1924.

A RADIO FOR PRESIDENT COOLIDGE, 1924.

RADIO SOCIOLOGY

WE HAVE always been convinced of the ability of radio to create a picture far exceeding that of television. In the latter, one's vision is only 21 inches across. Everything is laid out for the senses, and there's no chance for the game of unreality to creep in. We like to remember that good radio, with a word or an effect, can create a world in the imagination that is at once unreal and yet intensely personal.

What started us on all this was the series that we are doing on "The American Future" with David Riesman. (When talks like this come up, we want to set off klaxons and bombs to get people to listen—since our game is that we are alien to all these carryings-on, all we can do is program them at a good time and hope that people listen).

Riesman and his vision seem to us so real, so here. He's talking about the guy across the street who gardens every Sunday, whether he likes it or not, because his neighbors expect him to garden; he's talking about the factory down the way, with its conspicuous production—and the fact that people in this country do not resent conspicuous production half as much in war-machines as they do in education. Riesman marshals so many facts from so many sources: from historical writings, current drama, magazines, novels and songs, other sociologists and obscure government fact-books and, we have no doubt, from telephone books and nudie magazines. And one cannot help but be overwhelmed—whether one agrees with his conclusion or not—by his discipline with these facts, and above all, with his almost novelistic vision.

It's a favorite occupation of ours to think of the quasi-sociologists of the Riesman school as performing for the 20th Century what the novelists did for the 19th. Dickens and his enormous panorama of distorted characters serving to compile criticism of the social institutions of the 19th Century—a criticism of implication and juxtaposition. And—because the 20th Century novelists have moved on from the tactile world into the great, all-encumbering shade of Freud, the exterior social criticism must come from the world of fact reporting: the exquisitely detailed 'Reporter At Large' articles in "The New Yorker," the exhaustive documentaries done by some networks and the BBC, and the scholastic reportings of the Riesmans.

Before the Blue Eye of Television came to haunt us all, radio was the grandest source of unreality: 'The Green Hornet,' 'Allen's Alley,' 'Superman' made it possible for the great inward eye to produce a wild universe. And now, with fantasy in other hunting grounds, radio seems to be, increasingly, a view on what is. (We won't answer for the unreality of the teen-age love warblings which, if they are taken seriously, seem to set up rather grotesque ideals for future adults to build a world on.) Radio news, documentaries, Monitor going everywhere and doing everything—radio seems to have left fantasy behind.

And we always have to come back to the important role of the seers like Riesman and Leary and Greenson and Lomax. We often think of the plan of Lew Hill, the founder of the Pacifica stations. His ideal of broadcasting was to come on the air for only an hour—one hour in the evening, seven hours a week. But that hour would be a dilly, a real killer: that no one would soon forget. The plan has all the idealism of hope and all the reality of conciseness: and if we were to adopt that plan, it would be with "Suburban Sadness" or "Beating the Game" or "The Tale of Three Cities" or "Emotional Involvement."

March 1964

RADIO EGO

IT MUST have been Kirkegaard who said, "The adulation of ten voices can turn a fool into a god." Sometimes we are confused—for every now and then there are enough voices of praise around us to make us think that we might be able to survive by sheer ego alone.

We once chided one of our volunteers for thinking KRAB too important; he threatened to take all the fun out of it, for he was convinced that our secret audience numbered in the hundreds of thousands and that there were many plastic minds being molded this way and that by our different commentators. Yet if we were to believe his projection of numbers it would become such a burden we would have to go respectable to reflect properly the image others had of us; we could no longer be a cracked mirror for ourselves but would instead become an influential, dull, and pompous bas-relief for others.

Radio does tend to confuse one: all those kilowatts pounding in on the brain tend to warp it. The other day we had a rare burst of sanity when we watched one of our announcers on the air. He was in the other room so we couldn't hear his words;

but through the reflections of the plate-glass window and our own thoughts we could see his mouth move, even watch him nod and laugh. And we could not hear a word. "What a joke," we thought: "What a joke if there was no one out there listening, no one at all." Then he would be talking to himself, entirely fooled by the moving meters into thinking that he was not alone. What a joke.

One of our program participants described for us his feeling when recording a program, and then when hearing himself on the air. Recording ain't such hot potatoes: it's cramped and dirty here, and things are always falling down, and the words come out all wrong. But at home ("That voice! My voice! That is me!") there he is, coming out of that tiny radio and knowing that there are other tiny radios in other tiny rooms and thinking that his voice may be echoing over several thousand tympanic membranes—all at once. He liked that.

It is cramped and dirty here at KRAB: our main studio is our office and storeroom, serves as a library for 2500 records and 196,000 feet of tape; our hot-plate and tea-pot make it a kitchen, our books a study, our pillow a bed-room (one of our early-rising commentators came in without knocking and almost stepped on our face.) Sometimes the papers alone threaten to choke us—but then, there is a difference; during the day, a room and four walls until we push the red button marked ON and then the walls fall away and here we are on our naked platform and the only walls we have left are those in the mind and those others we drew on a map for the FCC—walls touching Rainier on one side and brushing the finger of Victoria on the other. We are there, of course—in all places between those walls, playing different roles for different people: putting some to sleep and waking others up.

It's like life, of course (a new game called Existential: finish the sentence "Life is a . . ." e.g., "Life is a flute that never got played" "Life is a mug-wump" "Life is a million-watt bulb: burned out"). There you are in your tawdry cell with the roof falling in—and then someone pushes the red button and for awhile you frown and nod and smile and talk, amusing some people, irritating others, boring still others—maybe even heard by no one; and then the guy comes along and pushes the off-button and the lights go off and the meters go dead and there we are with the dark walls back around us again and us wondering what all that laughing and carrying-on were about but not wondering too much because someone pushed the button and it's getting rather cold.

March 1964

JACK STRAW

"So hydous was the noyse, a, benedicitee!
Certes, he Jakke Straw and his meynee
Ne made nevere shoutes half so shrille
Whan that they wolden any Flemyng kille,
As thilke day was maad upon the fox.
Of bras they broghten bemes, and of box,
Of horn, of boon, in whiche they blewe and powped,
And therwithal they skriked and they howped,
It semed as that hevene sholde falle."

. . .*The Nun's Priest's Tale.*

ON WEDNESDAY March 26, 1964, the Federal Communications Commission granted transfer of control of KRAB to The Jack Straw Memorial Foundation, a non-profit corporation established in the State of Washington for just the purpose of operating such a broadcast station. The Foundation is made up of the various staff and volunteers of KRAB, and in effect gives legal clothing to what has been a fact since we went on the air; that KRAB is a non-commercial operation, broadcasting educational, aesthetic, and cultural programs on a non-profit basis.

With this action, we can now work towards getting official Internal Revenue Service recognition of the Foundation and the station as non-profit operations, which will then make contributions to the broadcast effort tax-deductible; this will open the door to appeals to foundations for improving the facilities and program opportunities of the station and—indeed—may make it possible for KRAB to operate without the devilish deficit that has hampered us in the past.

However, it is only the first step—for it usually takes the IRS a year or more to make such a decision.

The name Jack Straw has several appeals for us. Naturally, we delight in the obscurity of it. It refers to a trouble-making peasant type who, in 1381, led a riot against the Flemish inhabitants of London for nothing less mundane than economic reasons; but, better, is associated in Chaucer with the absolute confusion and demi-philosophical statements of Chauntecleer and Dame Pertelote under attack of the fox who, in turn, was under attack of the entire farmyard of the 'povre wydwe, somdeel stape in age.' Figure that one out.

Jack Straw bodes good for KRAB; with outside help we may be able to escape the inordinate confusion of our farmyard studios. We are sometimes revolted by our poverty and dream—as we have said—of glistening studios with miracle equipment and a transmitter lost somewhere in the clouds of faultless transmission and wild, improbable plans. We will refuse, of course, adamantly, to give up the confusion of our quasi-philosophical stance—that is the nature of KRAB and Dame Pertelote.

April 1964

RECEIVING RADIO-TELEGRAPH, 1925.

CHIEF LITTLE BEAR LISTENING TO THE RADIO.

TITLE MANIA

WE ARE rather easy-going with our titles at KRAB. When one of our volunteers is typing a letter, they always say "How shall I sign it?" And we say "Well, it's Thursday, so why don't you call yourself 'Corresponding Secretary,' but no, the moon is in the Gibbous Phase, so why don't you title yourself 'Special Events Director;' yet now that I think of it, Venus and Pluto are in conjunction, so I guess 'President' should do." Titles do give our stationery, tattered and dirty as it is, a bit of class, and it always improves and hastens the response, and it gives us a noble feeling to hand out a title that can be changed at will the next day. (In Broadcasting Magazine "Yearbook," which lists all the broadcast stations in the United States and their staff, we find ourselves stuffing in names like mad—names of people we haven't heard of for years: we like having as fat an entry as some of the television stations.)

Perhaps four times a week, there is a telephone call, and all the time the lady wants to speak to the Public Service Director; we quickly appoint one, and give the call to him or her. They can have it. What it is that these people want is a spot announcement read over the air . . . for the Seattle Symphony or the West Seattle Garden League or the Seattle Big Junior Little Great League Society.

A licensee of a radio or television station has responsibilities to the public—to enlighten, to inform on events and speeches and music and conferences that are taking place in a community. But if you are a big thumping hand-on-the-rating and heart-in-the-wallet type operation, you don't want to rock the boat and inform *too* much so you avail yourself of the ten-fifteen-or-thirty-second non-commercial spot announcement . . . sprinkled around when you are sure no one is listening. That's why you'll hear a great number of cancer or Radio Free Europe or Join-em-Up-Army-Bang announcements at six in the morning or after midnight—broadcast stations are discharging their public service obligations, and in such a way that it will not disturb their commercial announcements at 'prime time.'

Whenever these people call up asking for the Public Service Director, we explain that we appreciate their calling, but that as far as we can see, all the programming that we put on the air is 'public service,' that if what they are calling about is interesting or controversial enough, we will do a two or three hour discussion on it, and that we are not especially eager to demean their cause by giving it only thirty seconds. The best time we had was when the Marine Recruiter called up from down in the University Area and told us that he was sending up some recorded spot announcements, and we thanked him and told him that we had a much better way of dealing with the subject of Marine Recruitment: 'If you would care to come up, I think we could throw together a panel discussion on Recruitment and Recruitment Techniques.' He was delighted, and said that he would get in contact with his Commanding Officer. The next day he called back, somewhat more doubtful, and asked exactly what sort of program we had in mind. 'O, I guess we could look into the relationship between war, military training, and freedom,' we said, 'Or better still, we could get you together on a panel with a couple of people from the American Friends Service Committee—you know, asking some questions about how the services deal with conscientious objectors and the like; and maybe we could ask some questions about the real meaning of war—what it all means—you know.'

We almost had a program there—almost had a hot one; but the military services dithered around for a few days, there were some calls back and forth, and then we got a call saying that since they had a big week coming up, maybe they should call us in a month or so when they were less busy. That was all we heard.

Mebbe that's why we dislike thirty-second announcements: like everything in this instant, pre-digested society of ours, they give a man no chance to answer back, to question, to retaliate. Truth may come in small packages, but the crumbs keep leaking out.

April 1964

RADIO FANCY

"I HAVE to leave the radio on," said a younger friend of ours—"tuned to KRAB, even when I am not in the room. Because I have to let all those voices out of the box: all those voices arguing and discussing and commenting and explaining—they have to go somewhere. They just can't stay trapped in there. Besides, even when I am not in the room, if the radio is on, the molecules in the walls and floors and things are being changed; the molecules are sopping up some of the vitality of all that dissent. It's all very important for the atmosphere of my room."

As far as we can see, the best thing this next generation has to offer—if a generation has to offer anything—is a fine sense of fancy. Sort of a combination of Lewis Carroll-A. A. Milne-J. D. Salinger-and Nathanael West. We need something: the elimination of so much of the poverty and its attendant hunger and discomfort, coupled with the easily available education and television-induced sense of adulthood, leave the younger people we know faced with themselves at such an early age. All the 20 and 22 and 24 year-olds we know are haunted by that big kaleidoscopic mirror long before they should be forced to; and we feel you can look into your own eye for only so long, and then the dark sea somewhere there behind the pupil begins to haunt you. Then the only relief is in drink, death, or fancy.

We have indulged in our own sense of fancy from time to time; we have spoken of the waves of radio passing through clouds and trees and cows and things, tickling the ear of someone perhaps in Sequim or out on Vashon; we have tried to describe the walls that fall away from us at 5:30 or 6 every evening; and we have continued to dream of our own particular Isle of Gold, where lights blink on at our very step, and the equipment hums obediently and properly, instead of growling, smoking, or falling apart as our present equipment tends to do.

"To see what I see," said another younger friend of ours, "you have to be a Scott FM tuner when you grow up; your head has to be full of electronics—you open your mouth, and there is a row of transistors." (Shades of Capek, we thought.) "But see that tuner: it's dark, you have gone to bed, there is no sound around, nor light—except the single light of that magic eye. And you turn the knob, and when you pass over the frequency of a station, from an open, incomplete circle, it turns to a ring-of-Saturn (or an eye that winks at you knowingly); and you find yourself trapped in a bright green band of transmission."

We must admit we got lost with the mouth full of transistors: we kept thinking of the engineer types we have gotten to know: they have their own special type of magic transmission with other engineers ('Well I got the grid-dip up around the selinium diode trip-match, and wrapped the final phase circuit condenser under the rectifier modulation shift-inverter, then I told him what to do with his plate-revolter circuit . . .'): we keep wondering if the engineer types visualize their kidneys as little condensers, their eyes as orthicon-image tubes, and their hearts as some sort of complicated push-pull amplifiers.

"So you've got to see the signal, first of all," he went on. "You know the station is there by that green eye that snaps closed. And . . . and . . . what it is: each broadcast station all across that band: it's a big room. A giant room, with many voices in it.

Sometimes the room is good: you go in and you want to stay, and the door closes. Alan Watts is talking all around you, or John Cage is coming in on all sides. (And sometimes the room is bad—you know: they keep trying to sell you Preparation H or a Ford.) But if the room is good, and the door is closed: sometimes, after the station goes away, and the green walls are not really there, and the floor has sort of fallen away: it stays with you. That voice is still going through you, the music is inside you: long after the room is gone."

May 1964

ENGINEERS

THESE engineering types are always playing games with reality. The other day, one of the mad ones in that department (we speak as if KRAB had a veritable army of engineers, whereas we have only two—but the things they do with those wires and tapes and things convinces us that they are mad. All of them.) made some magic with a light-bulb and two pieces of wire. How it works is this: take a small red light-bulb and connect two wires to the base of it. These two wires are about two feet long and connect to nothing, as far as we can see: but if you set the whole thing on the fence just below our antenna, then whenever the transmitter is on, the light-bulb glows.

The Engineering Type tried to explain it to us: there was some fol-de-rol about R F Currents and Induction and Force Field—but we know better. We know it was magic: the Great God of Radio has been bribed somehow to turn on the light at the right time.

Those who have anything to do with broadcasting always seem to forget the magic. They think it's perfectly normal that the voice could be thrown for hundreds or thousands of miles with no visible support; they see nothing strange about projecting the entire *I Musici* over to Bremerton or Victoria in an instant. That's old stuff. Yet we try to resuscitate the wonder that guy felt there in Wichita, back in 1920, huddled in the barn over his crystal set, excited with the earphones, hearing KDKA (all the way from Pittsburgh!), hearing jazz (wow!), hearing it so clearly that . . . why it might be right there in the room with him. What miracles!

As time goes on and electronics more perfect, the magic gets more ominous. We've seen, for instance, a nightly raising of the dead—every time we play a tape of a talk by Aldous Huxley or that record of W. C. Fields, we've brought them back to life. And that dreadful weekend of the Assassination, it was very hard to believe that Kennedy had died—we saw him revived a thousand times that weekend by means of film—saw him laugh and talk and smile. It's all a mad juxtaposition of different times which, by coming through the same TV screen, are very hard to separate.

We see the magic as ominous in—for instance—the television newscasts. We get five minutes of international murder and mayhem, then quick dissolve to the May Queen, unpleasant and human because of Upset Stomach; then dissolve again to a mechanical picture of her stomach full of evil acids; then a dose of Tums and dissolve to the May Queen with an army of followers and her stomach right! right! again. This juxtaposition of mayhem and miracle of Tums leads to all sorts of irreverent questions: if Johnson used Gleem would our foreign policy come all right again; if Khruschev used 4-Day Deodorant Pads would he and Mao be able to make up?

We have a theory that people who work around radio too long get soft in the head: it's all them kilowatts pounding in on the brain. Broadcaster's syndrome is characterized by a great belief in words and their effect—as if everything that is transmitted becomes important solely for that fact. This leads to a great confusion: not only of dream and reality, but that—say, the Beatles and Sani-Flush are just as important as an exploding population, or the bomb. And who can blame them: through exposure, the commercials become more real than the mystery of human social behavior.

May 1964

STORM TROOPERS

KRAB was plunked down, almost by chance, in a part of Seattle where lawns are treated almost as lovingly as Motherhood, Chastity, and the TV set. There is no small concern with the dishevelled appearance of our building, the all-hours comings and goings of dishevelled program participants, and our dishevelled engineer's weekly journeys up the pole to bang on the antenna and look at the view.

The only redeeming features of the locale are Julie's Cafe (where we get our daily cup of coffee and weekly lecture on our unkempt appearance) and—pardon the name—the Speed-ee Mart (where we get our daily bottle of beer and box of Animal Crackers.)

It was during our trip for the beer that we passed the old lady in the wheel-chair, sunning herself and reading her checkbook, and she said "help." One never knows what to do, in this age of bombs and violence, when someone says "help," so we stopped and smiled slightly and pretended we had stopped to think an Important Thought, and she said "Help. I'm being held against my will. Call the police. It's my family: I have an apartment, but they won't let me free, and help, I have $13,000 in the bank, and I'm a prisoner, and all I want to do is be alone in my apartment. Help."

Last Sunday morning we went down to the Hyatt House to do an interview. One of our volunteers had called and said "Hey . . . do you know Norman Rockwell is here? Are you going to get an interview?" And we thought: "This'll be fun: we've been waiting a long time to ask someone about those wholesome Thanksgiving Scenes on the cover of The Saturday Evening Post" so we called up and found out he was wrong. It wasn't Thanksgiving at all. It was Storm Troopers and Concentration Camps and Jewish Traitors—a somewhat different Rockwell, to say the least.

In better times, we would have given up the interview and just thought for awhile about over-romanticized Motherhood and Naziism and drawn some parallels and slept late on Sunday; but it was bright and cheerful that Sunday morning so we went on down and interviewed this other Rockwell: sat in his prison, the motel room, for awhile and talked about all sorts of garish things that some men can cook up, and recorded it all and then went away and thought "Help. There must be something significant in all this," but there wasn't, of course, only a devilish stomach-ache.

We have said before that all men create their prisons, and we guess we prefer ours: somewhat dirty and ill-kept, but aesthetic anyway, with beautiful bars called Truth and those nice chains called freedom-of-expression to caress; yeah, we like our jail just fine, but with all these others, we can't avoid calling 'help' now and then.

June 1964

SOUND EFFECTS MEN CREATING A COLLISION, 1930.

MR. AND MRS. MAVIS PINGRAY, INDIANA, 1938.

WHO IS TO HEAR?

"IN 1917 when you said something strange or subversive, they put you in jail. Nowadays, they don't even listen to you . . . you can talk until you're blue for all they care." So said Ammon Hennacy when we interviewed him last year. He ought to know—in his 47 years as a Catholic-Pacifist-Anarchist, he has been in jail 35 times; "but," he says somewhat wistfully, "not lately."

We are rather ambivalent here at KRAB—about listeners and being heard and all those ego-filling things. When we put a great deal of time and work in on a program, we cannot imagine it being broadcast and not being heard by a great number of persons. It is essentially illogical, but when we spend four hours driving to, taping, bringing back, and editing a twenty-five minute speech; when a commentator spends sixteen hours preparing a thirty minute talk; when we have a music program shipped from half way across the world; when we spend eight hours on the telephone for an hour-and-a-half panel discussion: then we cannot comprehend no one out there: listening, reacting, being changed.

But at the same time, we have often contemplated the

possibility of broadcasting to no one. One time, when one of our applications for broadcast facilities got stuck in the black craw of the Federal Communications Commission, we thought of asking for a station in the swamps of Okeechobee so we could broadcast to the alligators and possums and ourselves. "One of the best things about KRAB," one of our volunteers said recently, "is that I often get the feeling—with your two days of music from India, or your plays in Arabic, or your Rexroth book reviews—that you don't really care whether there is anyone out there at all. That's why I listen: so I can be part of the great uncared-for, the great unwanted, the great nothing of everyone."

Once, for no apparent reason except a slight case of Broadcaster's Syndrome, we stayed on the air past midnight for five or six hours, playing Bach Cantatas. "There," we thought . . . "that got rid of the last of them: no-one can stand 'Ich Habe Genug' at 4 AM." Yet at six, when we yawned and turned off the transmitter, and were falling out the door, the telephone rang: " . . . baroque music before breakfast . . . all night . . . how wonderful . . ."

So there we are; broadcasting to everyone and no one at the same time: stuck with two philosophies at once opposing and compatible, the Yin and Yang, Prometheus and the rock, Miss Lonelyhearts and Shrike, Daedalus and the sun, Archy and Mehitabel.

June 1964

REGULAR CHAOS

IT WAS another dull week at KRAB. The tape we get from the Italian Broadcasting System has the electronic music wound backwards and at the wrong speed—we play it this way, and a listener calls to praise the music as 'interesting and creative.' We lose our master program guide for the next two weeks, spend hours going through all those dreadful old letters we never answered to find it, and ask ourselves whether Deb Das dropped it into his mound of papers by mistake, or whether the after-hours drama group got it mixed up in their scripts, or whether the Morning Man, in his charming compulsive way, threw it away because of 'all these damn papers sabotaging my sense of Order.'

Four commentators in one week fail to turn up—a new record. The transmitter looks poorly, causing the installation of a rope (with a noose on the end, of course) which goes up there heaven-ward somewhere into the bowels of the antenna; 'when the line current goes below 9.5, pull on this,' says our engineer (and, we would imagine, in real times of crisis, we should simply hang ourselves from it).

Through some electronic vagary which we have yet

to adduce, our African Periodicals program goes on the air with a background of a meowing cat—which gets louder and louder and finally, to our extreme discomfort—inundates the program entirely. During the Stravinsky Birthday Concert, the jazz director, the music director, and a staff member join in a spontaneous 12-tone rendering of 'Happy Birthday, Dear Igor.' We get a riotous tape from a recording laboratory in Californa, but some of the language is indelicate and we have to hide the tape for fear some volunteer will play it by mistake. A lady calls up to ask if we will advertise over the air that she wants to sell her German Shepherd—we go on and on with a long philosophical discussion about the nature of radio, the nature of being a dog, the nature of being a dog on the radio, the nature of being a radio on a dog, and finally wind up by telling her that we have a German Shepherd that we simply have to get rid of and asking whether she will advertise it over the air for us. She doesn't seem confused, only grumpy.

We sometimes wonder how the listeners can stand it; we sometimes wonder how *we* can stand it. If it's happened one time it's happened a thousand times: someone will come to us and praise us for our courage in starting KRAB and our bravery in maintaining it on the air. This always bothers us—because we couldn't think of any other way to spend our lives which will grant us the chance to involve ourselves in the life of a community and perhaps in some way to even influence the course of that community. Actually, given the continually tedious finances of KRAB, we would advise our well-wishers not to confuse courage with sheer pigheadedness, since we find the latter a much more viable commodity in this business than the former.

The picture of us marching valiantly forward, stamping on the snakes marked 'ignorance', 'prejudice', 'intolerance', and 'evil', with 'Excelsior!' tattooed on our chests and '*Vae Victis*' writ large on our less seemly parts, is a picture that no doubt appeals to us; but we like the other face better: the dear testy lady, trying so hard to find a proper home for her German Shepherd, and bedevilled by these idots who claim they operate a radio station, but just seem bent in twisting up all her words and her careful logic into an unseemly mess. Pshaw.

July 1964

FANFARES

THERE'S a whole different approach to life and to literature that seems to appeal to a certain minority—sort of a droll, self-depreciating, amusement-in-tragedy approach which, before WW II was represented by Brecht, Mencken, West; it traced back directly to Swift's 'Modest Proposal' which is more or less the introduction to the whole school. It was given a boost by the war and the Jewish 'Ghetto Humor,' and has recently flowered under the aegis of Günter Grass, Joseph Heller, Terry Southern, popped up in Dr. Strangelove (much to the discomfort of critics), appeared in the guise of "The Realist" and "Mad" magazines, and is seen in its most bitter form in the monologues of Lenny Bruce and Dick Gregory. We see it as a refusal to mourn the decay of the whole human race: to be somewhat amused by it but, since we are all stuck with being humans, not to be involved in the tragedy of it all.

Unless one is awfully dull, one cannot help but get involved in this tragi-comic aspect of life. We often cite the strange juxtapositions on television: a picture of the slums of New York, a burning village in Viet Nam, an army of starving refugees in Hong

Kong, followed immediately by the Playtex girl, dancing across the screen, happy in her living girdle, smiling as if she and her girdle didn't have a care in the world. Since most people fight to forget the continuity in life, they are able to ignore this odd intermingling of poverty and wealth, prosperity and desperation. It's the difference between the average television viewer, slouched low in his Bide-a-Way, Pop-Up, Comfy-Bed, his instant Lobster-Newburg TV dinner in hand; munching on the plastic shells and blearily eyeing the high life on The Ed Sullivan show—it's the difference between this sort of mental hibernation and the Heller or the Grass, slouched low on the bed of life and blearily munching on the bones of man, eyeing his whole sorrowful condition.

It is in the self-depreciation that we try to be closest to the tragi-comics . . . in our announcements, in our introductions to programs. We never try to lose sight of the fact that this is a small station in a small corner of the United States with a small audience. We often refer on the air to "our ill-printed program guide," or "our crowded and unsightly studio facilities." There are many broadcast stations as unsightly as our own, but you'd think, the way they talk on the air, that they were NBC—it's all a matter of the image.

The troublesome thing of course is that people tend to get deceived by our way: because we do not introduce our programs with fanfares and take them out with canned laughter, they can't be important. People have been deceived and shouted at for so long that they cannot believe that understatement can lead to anything vital. We could easily imagine how a commercial station would present our commentary series:

"And now (trumpets, leading up to a fanfare by chorus of angels) KRAB, your voice of news and views, your station with (applause) *dedication*, presents (drum roll of thousands) George Glutz (cymbals) noted author and scholar, with (crash) 'Commentary.' (Martial music, fading into whisper of angelic chorus.)"

Of course (what broadcasters in general don't seem to realize) there is another virtue in underplaying presentations; if George Glutz turns out a whiz-bang commentary, then we can feel proud for not having overplayed him. However, if George Glutz turns out to be a fig, then there is not the shame of a ridiculous juxtaposition of fanfares and things with some cold lifeless words and cold lifeless thoughts.

July 1964

GRU

ON THE island of Drobb (or 'Man's Hope') the great grey philosopher Gru lies on the sands of the beach scratching in the sands with a twisted stick. The natives of the island of Drobb know enough to leave the great grey philosopher alone because it is rumored that he is very wise although he salivates a lot and has a tendency towards fits (the shrieks mingle with the waves and are heard in every part of the island for Gru can be interpreted to mean 'philosopher-of-the-big-voice'). It has been said that the scratchings in the sand, although unintelligible to the natives, are in reality the face of God. Each day the scratchings are different and each day the natives say that Gru is putting a new face on God because each day the course of humanity changes from the day before. Somedays if Gru puts a good face on God there is prosperity somewhere in the world and somedays if Gru puts a bad face on God then there is catastrophe somewhere in the world. And each day when the sun falls into the sea to drown the moon comes up to tug the seas over the loin of land called Man's Hope and the waves fall on the scratchings and the face of God is erased.

But always the philosopher Gru has been fed and clothed by the natives of Drobb even though he is cranky and cantankerous and has fits, for even the natives of Drobb realize their responsibility to the rest of the world and the necessity that the face of God be scratched in the sand of their beach.

One day the Men of Civilization came to the island of Drobb in a big boat that spouted steam and fire. The Men of Civilization wore white uniforms with gold braids and white caps with gold braids and white shoes without gold braids. They walked carefully so as to keep their white shoes white and carried swaggersticks and when they heard of the philosopher Gru they walked to the beach carefully and found him there lying in the mud with a stick in his hands. Gru was naked and scabrous and wrinkled and old and salivated a great deal, so the Men of Civilization averted their eyes to look at the palm-trees or the sea which was colorful but colored no better nor worse than the sea of any paradise. The leader of the Men of Civilization was a brave man and had seen dirt before and had heard a great deal about the power of Gru for the benefit of mankind so he took his courage in hand, wiped some saliva off his foot, and said to Gru: "Sir: I am a man who comes from a mighty civilization; we have large edifices in which we work and live, we have food and prosperity, we have things that roll on wheels that carry us anywhere, and we can even throw voices and pictures over thousands of miles. And yet, for all our wealth and buildings and food, something seems to be missing . . . something is wrong: it is that still in our world there is fear and hate and . . . and . . ."

The natives hidden behind the trees watching frowned and scratched themselves. No one had dared to speak to the philosopher Gru this way (indeed no one had ever dared speak to the philosopher Gru because it was rumored that he was deaf and dumb) and some of them thought that the old man would throw a terrible fit and shriek and scare all the animals and all the children, and others thought that perhaps the philosopher would scratch a good face in the sand so that the Men of Civilization would return to their land and find that there was prosperity and happiness. But the philosopher did neither and just lay there in the sand coughing and spitting and finally the Men of Civilization went away and never came back.

The island of Drobb (or 'Man's Hope') is quiet now for that night the philosopher Gru rose painfully from the heap of mud he called home and crept painfully into the waves, his body crashing into the waves until finally his wrinkled bald pate disappeared under the waves without even a bubble and sometimes the natives wonder if things are so empty now because there is no one to lie on the beach and growl and screech and take a bent stick and scratch little lines in the sand (which they were sure were the lines of God's face).

August 1964

BROADCASTING FROM THE KITCHEN
OF THE SHERMAN HOTEL, 1929.

MADAME ASTA SOWORINA AND HER SONS
LISTENING TO THE RADIO, 1922.

KDNA

I WAS on the air, playing some Gagaku Court Music of Japan. A listener came in. He was wearing a cape and a magic pointy hat. He gave me an eggplant.

"Here," he said: "Take it. Do you know that this eggplant contains all the vibrations of the Universe?"

I didn't, but I took it anyway. I set it on the Gates Studio Consolette to keep me company as I was playing the Japanese Court Music.

Now I had never been exposed to the Vibrations of the Universe before. At least I didn't think I had. But since I was an old radio man, I figured I could hear them by simply holding the eggplant up to my ear.

I wasn't sure of what to listen for. Once I had read in a book that the Universe turned on slots of gold. I thought that the noise of this particular eternity might be a soft, golden moan.

I put the eggplant up to my ear and as far as I could tell, there was no especial sound. At least *I* couldn't hear it. All I could hear was the Gagaku Court song "Ichi Uta."

After the selection was over, I told the KDNA audience about this adventure. I said: "Here. Maybe my hearing is bad. Or maybe I am just insensitive. Maybe you can hear the sound of the Universe better than I can." So I held the eggplant up to the microphone so that they could hear what I couldn't hear.

And as I did that, the tape machine (which I had neglected to shut off) came on with another selection of Japanese Court Music "Suruga Uta."

At KDNA, the radio station and the volunteer living quarters are all in the same building. When you get tired of playing radio downstairs, you can go up to the 2nd floor and borrow one of the beds for awhile and wake up at 3am and hear Gabriel playing James Brown at the 120 db level. . . and it creeps up from downstairs through the heat ducts like fog, mixing with your dreams, and outside the Blacks come and go all night at the Rex Hotel & Bar next door.

KDNA is set in the St. Louis ghetto. The streets are filled with pimps and whores and angry Blacks and drunks and kids in rags playing and forty-five-year-old bohemians and junkies: right outside the window you can see them laughing and talking and running and falling down.

(The neighbors are convinced that KDNA is a whorehouse in disguise. Two drunks came in once, wanted to see the girls. "No," Jeremy said: "It's a radio station. Look." And he took them back to see the records going around, the sound coming out. They weren't convinced. So he went on the air, said that there were two visitors who believed themselves to be in a house of prostitution. He asked that some listeners call up to explain that it was the Sound of Music, not young ladies, that made that particular operation go around. They did, but there is no record that the visitors were at all convinced. Prejudices die hard in the ghetto.)

The last time I was at KDNA, I was drafted to do all the shitwork on preparing their application for renewal of broadcast license since I am such a good bureaucrat. Besides, I am the only one connected with that operation who can spell.

One of the questions on the form really got to me. It

said *State in Exhibit Number (blank) the methods used by the applicant to ascertain the needs and interests of the public served by the station . . .*

"Jesus Christ, FCC," I thought: "The needs of St. Louis come right in the goddamn door and steal the tape recorders and tools for another fix. And when they aren't doing that, they are following the female volunteers down the street, trying to get them into the alleys so they can expose them to other community needs." It's just like asking a concentration camp inmate fresh out of Auschwitz whether he was *sure* the Germans were prejudiced against Jews.

As far as I know, KDNA is the only ghetto radio (or television) station in the country. I must say it adds a certain *verite* to the programming. Like the staff has to go no more than twenty feet to see a city dying because of ruinous speculation, and a petty, bickering city government, and prejudice. But living and working in the ghetto makes the staff tough, less than willing to be open and free as we are (I think) in the garden paradise flower of Los Gatos. They have lost too much equipment to the sticky-fingered junkies; the women have been too brutalized on the empty streets. There are two locks on the front door; a peep-hole and speaker, and when you knock after six in the evening, a 1984 voice says "Who's there?" *That Orwellian voice is one of us.*

Across the street, the Apollo Broadcasting Company bought up a two-story building and ran an automated FM station there until the ghetto kids burned it up for the third time so that they got sick and tired of it all and moved downtown to the marble-and-glass section of St. Louis where the needs of the people weren't so *pressing* and *real* for god's sakes. KDNA staff shrugged their shoulders and boarded up all their street-level windows and bought seven more fire extinguishers.

When I was there, I realized that KDNA sees so many community needs that they might well become one. Radicalization, I guess they call it. What happens if you are a young white college dropout hippie type, into experimenting with good radio—and you find yourself in the middle of the dingy community of the poor—with all its accumulated grievances: it's enough to make you think, "As part of the most gorgeously rich country in the world, why does this city look something like 1944 London?" Are all Blacks just slobs who tear down neighborhoods and rip up buildings? Or is it something more subtle than that? *Like the destruction of self-belief of a whole culture—born of slavery, prejudice, five centuries of intolerance?*

So you find your ideas becoming a bit more militant. And then there's the St. Louis Police. Doing their bit. To radicalize.

It's not enough for the staff to be living pisspoor, getting $10 a month for expenses, giving up every possible freedom of affluence to this crazy station—then there's the force of order and law.

It is a balmy night in September. Late at night. 1:18 AM, to be exact. And they come bludgeoning in the door, dragging the staff out of bed, claiming to have found a couple of baggies of you-know-what (rabbit-and-hat style) in some previously empty cabinets.

KDNA, a station which has programmed some material which is very critical of our foreign policy, state and city governmental corruption. A station which is striving to exercise those all-too-rare freedoms of speech. Is that why you are here, officer? I would like to explain to you officer, but there you are, waving the two black eyes of a sawed-off shotgun at me, warning me not to tell the listeners about what is happening to me, and my heart, and the future of free speech radio in St. Louis. I worry about your finger, officer, and blood *my blood* and flesh *my flesh* all over this quiet control room. Are you trying to radicalize me, officer? Are you trying to give me some feeling for arbitrary, untrammeled power, police power? Is that the reason you are here, officer?†

Sex and Broadcasting 1974

†Jeremy Lansman, the manager of KDNA, once claimed that "someone was out to get the station." I relegated this to my *Community Radio People Are All Overwrought* file. This despite several police raids, detectives seen following on-the-air commentators from station to their homes, and a series of scandalous front-page articles in the "St. Louis Globe-Democrat," which placed KDNA somewhat to the left of the 13th International Congress. Then, in 1980 or thereabouts—I read an article in "The Wall Street Journal" which stated that many of the Richard Nixon/John Mitchell Counter-IntelPro activities had been concentrated in St. Louis, and that many legitimate dissident Missouri organizations had been infiltrated, with *agent provocateurs*, informers, and other demoralizing (and illegal) government operatives. I called Jeremy to apologize for scorning his suspicions as paranoiac. Of course, because of his ulcers that grew out of those years, my apology might have come a bit too late.

THE NATIONAL ASSOCIATION OF BROADCASTERS

THAT Vincent T. Wasilewski, President of the National Association of Broadcasters, looks just like a stuffed tomato. They cut off the top and stick it full of tuna and mayonnaise, and by the time you get to the table, it's old and sort of crusty. It's a dish dear to the hearts of caterers—like those who took care of the National Association of Broadcasters in Las Vegas last month. And I am damned if I could tell the difference between Mr. Wasilewski and my lunch as he addressed us broadcasters and gave us that old song-and-dance about The First Amendment and our inexorable right to make a mint because of it.

I was thinking—what with those constant references to our Service to the Public, and the Public Weal, and such—that broadcasters aren't too different from the oil companies. We have stations with a variety of services; we coin money with our towers; and every time someone questions our right to make a 39% net return on invested dollar with the kind assistance of the U.S. tax laws—we start talking about Free Enterprise, and the Bill of Rights, and Capitalism vs. Other Governmental Systems; and through some spaghetti logic which I will never ever comprehend, we equate Freedom of Speech with coining gold and—whenever some of our unfortunate brothers in the industry get out of hand, we hire a Doyle Dane or J. Walter Thompson or Vincent T. Wasilewski to set the record straight again: public service, benefits to society, freedom—a whole jugfull of good things that will come our way as long as those creepy Nader-types stay off our backs, stop threatening the American Way.

At the Convention this year—Las Vegas, April 6-9—they gave out the usual awards. They ignored Nick Johnson and Phil Jacklin and Al Kramer: being bodily lifted out of the sweat-shop mentality of the 19th Century is not something that gets you awards at the NAB. No—they honored Jack Benny and George B. Storer. George B. Storer! I found making an award to Storer not unlike, say, striking a medal for Attila the Hun for his energetic Land Reform measures.

Storer was there, came up in front of the throng to get his prize. He didn't look like a tomato, and he damn sure didn't look like those rosy-cheek chin-thrust photographs that peer confidently out at you from "Broadcasting." Rather, he looked to me like one of those frail old men that sit on the benches at Hemming Park and try to show you pictures of their children from 40 years ago: splotchy old photographs of people you could never care to know—straight from a world that doesn't exist any longer. And Storer's speech was in that same vein. He talked about some flood in Akron that he covered 200 years ago, and how broadcasters were holding their own against the tide of Keynes—and he didn't mention a word, didn't breathe a sound about what he personally, George B. Storer, had done to vulgarize American broadcasting. And, bless me, he didn't mention those Uplift ads that his stations carry from time to time in the trade magazines—where they address themselves to the pressing problems of our time: crime in the streets, rape, governmental stupidity, burglary; nor indeed did he address himself to the part that Storer Broadcasting had played in raising the expectations of the poor and the hungry through miracle 60-second commercials: seeding the minds of the hopeless with defeatable hopes, so that they do (can't help but) indulge in crime in the streets, rape, *et al.*

Las Vegas was a perfect place for the NAB convention. All the sharks and gamblers and gaudy skirts from Los Angeles and Anaheim and Daly City were on hand. I got some sort of a feeling for the perfect marriage of those people coming together in commemoration of the easy dollar. Gambling in Las Vegas is the fruition of The American Dream: for little effort, one should be able to make a large fortune. In broadcasting—they have done it. In Las Vegas gambling—they are still trying.

Or maybe it's better to compare the licensees of the FCC with the licensees of the Nevada Gaming Commission. Only the very stupid lose their shirts. And they weigh the odds so that the chances of loss are zilch.

One of the most recent broadcaster's card-stacking devices is something called TARPAC, which sounds like a road-building compound but is in reality a heavy-duty lobbying organization. The literature of TARPAC states unashamedly that their job is "to take contributions from hundreds of broadcasters and channel them to candidates for the Senate and House who are friends of the industry." They state that they made "modest donations to 85 candidates, and there were more than one hundred others who asked for contributions from this committee." Which left me wondering if my local representative had tried to put the touch on this highly-financed organization; and, if he did, left me wondering even more about his independence.

The other congressional buy-and-sell organization is called The National Committee for the Support of Free Broadcasting, which interested me greatly because I am in favor of free broadcasting as much as the next man. It must have been Tallyrand who said that you can do anything if you put the right appellation on it—which is exactly why they put the "Arbeit Macht Frei" sign over the entrance to Auschwitz. I guess we can well understand why the NAB did not call it The National Committee for the Support of Closed Broadcasting, or The Gold Mine Perpetuation Commission.

If this year's meeting is any example, broadcasters are still banging the drum loudly for the 5-year renewal bill, but I'm beginning to get the feeling that some of these are wiser than that.

Or, as a very good friend of mine, a communications attorney, said: "If Harley Staffers or Torbert Macdonald hears some program he doesn't like on WRC the day they come over the Potomac to vote on a 5-year bill, broadcasters are going to be stuck with something ghastly, like government sanction on any and all strike applications. I'm advising my clients to drop the 5-year thing."

I asked one of the employees of the NAB about the representation of blacks or other minorities on their board. Turns out that of the 50-member board—one is minority and two (just recently elected) are women. Since President Ford was scheduled to speak on the second day of the convention, I had this brief fantasy of me standing up and asking him why he would consent to appear at a gathering (and thus put his imprimatur on it) which was so blatant in its disregard of the hopes and aspirations of the minorities of the country.

Fortunately I was relieved of the obligation for such an appearance by the onslaught of a spectacular hangover induced by some overdone "hospitality suite" the night before where, they tell me, I tried to dance on the table with a lampshade on my head. Just as well: On Wednesday, Rev. Jesse Jackson of PUSH delivered a 20-minute zonker speech, excoriating broadcasters, the FCC, all of us for being so slow to encourage minority ownership of radio and television stations. His speech was greeted by the polite applause of 1500 assembled broadcasters—reminding me of Karl Marx in modern day fantasy, where he would be invited to speak before a variety of service clubs like the Kiwanis and the Rotary; and the chairman, at the conclusion of the speech, would say, "Thank you, Dr. Marx, for a very provocative speech. And now I have this message from the Ladies' Nite Committee. . . ."

My friend Cruickshank stole my Ford speech ticket, stole it right from my nerveless morning-after fingers. He assured me later that I had missed nothing, what with my cerebral hemorrhage and all. The 200 Secret Servicemen and their pointy shoes had made all the waiters and waitresses leave the room before the President's appearance; and there were two gorillas perched on the roof of the hall with shotguns in case any of the audience failed to laugh at the prearranged jokes. Kissinger made a surprise appearance, although God knows what the NAB has to do with American Foreign Policy. And Cruickshank, who was trying hard to be L.W. Milam (he had borrowed my badge to get in) was terrified that the SS would demand some identification and put us both in the pokey for exchanging personalities without governmental permission.

▪ ▪ ▪

My friend the lawyer and I had a farewell dinner Tuesday night in the Kabuki Inn at the Hilton. I have a deep love for Las Vegas because the phoniness is so close to the surface that you never have to think that anyone's trying to put you on. Like the entertainment in this pseudo-Japanese outpost consisted of two Sansei from Oakland rampaging around the stage with black scarves around their waists, plunging and banging at each other with silver-painted plastic swords while fluorescent lightening zonked around overhead and warm tap water drizzled down in a wry simulation of rain. Neither this nor the shrimp dipped in library paste and labeled "Tempura" could fool us into thinking that we were anything but 17 light-years from Tokyo—but my lawyer friend, who (as is unfortunate with lawyers) is a stickler for detail and what he presumes to be reality, had to tell me of the most miserable weekend of his life spent a couple of years ago in Las Vegas.

Seems he was in town for a revocation proceeding against KORK. The hearings dragged on and went over the weekend. Everyone left town except my lawyer friend who cares little for drinking and none at all for gambling. He thought that he would find a park to sit in and read—but evidently all parks had been outlawed in Las Vegas. He then tried for a museum, but two wax figures of Jesse James and seven guns didn't remind him of the West, much less a museum. Finally he looked up in the Yellow Pages to find an antique shop. "That's what I'll do," he thought on this particular dull and feverish Saturday, "I'll buy some antiques."

Las Vegas's only antique shop is down at the end of the Strip—and from thence another three miles down a dusty road. "It wasn't the sweat and the heat that got me," he said, "nor was it the cadaverous old geezer with no teeth who ran the shop." Nor was it the display of coffins and spiderwebs, under the single 40-watt bulb. There was something more deathly than he had ever seen before. It had to do with the face of the man baked by 75 years of endless sun, and the bleached boards of his floors, and the stuffing leaking out of the old man's seat.

As he was telling me this, I kept thinking of George Storer—the old man of radio. I kept thinking that with his money and stories of floods in Akron in 1914 he was still no more than an old geezer in an antique shop with coffins and webs and too many yesterday artifacts: him in the bleak, dark, dusty antique shop of radio—with the rags of time settling around his shoulders, and the stuffed owls at the end of the road. God knows that he changed broadcasting in an outrageous way 25 years ago, feeding off the robber-baron mode of business and exploitation of resources. But his time is coming, has come; and those children and the poor who have been tantalized by his 6-second miracle $1000-a-minute messages over his $100,000,000 worth of radio and television stations are beginning to feel something besides acquiescence to a broadcast mogul's dream of giga-dollars and megaprofits. Expectations have been raised to bursting by the sixty-second cure, projected so winningly and so attractively to a hundred million homes—of which at least 20% are stuck below the poverty line. Several thousand politely applauding white anglo-saxon businessmen can give one a sense of power and impregnability—and when the President joins that number, you feel like you can never be shaken. But the natives are powerfully restless and the broadcasters have helped that restlessness. Too many poor and isolated and anguished have been tantalized with what they could have, but don't have. The old men in their antique shops nod their heads and bar the doors to keep out the riff-raff, but the nameless young and poor and angry and frustrated are coming to be heard, have to be heard; they have to come to know their power to seek out and overwhelm the foolish and the old who have dominated the aether too long, and who have no sense (they don't watch their own messages) of the changes that are at hand. Restlessness and anger never obtrude when the disadvantaged are at the bottom of the cycle; no—it is when the curve is starting to swing up, and they have a taste of what it's going to be like at the top, or over the other side. That is when the dispossessed become enraged and out of hand: that is when they rise up to seek out the wrinkled old men guarding the industry coffins and yesterday's pride; that is when the young and the ecstatic come to overturn the symbols and machinery of the dead and insensitive, and induce the change that we all know must come.

Reprinted from "Access Magazine,"
May 1975

BROADCASTING STATION OF
THE WOMEN'S ATHLETIC CLUB, 1930.

SKEEZIX STARING INTO THE SPEAKER
OF AN ATWATER KENT, 1926.

TV BLUES

WTOP-TV is a commercial television station located in the ninth largest market in the country, that being Washington, D.C. The station is licensed to *The Washington Post Company*, of honorable journalist fame. The difference between the newspaper and the television station is considerable, I do believe.

For instance, WTOP-TV is a governmentally-protected monopoly—*The Post* is not. The latter is subject to all the troubles and vagaries inherent in a freely competitive capitalistic democracy (or whatever-it-is system we are supposed to be living under this year). The television station is, on the other hand, firmly protected against any unnecessary competition by the full force, might, and bureaucratic power of the Federal Communications Commission. And if you have doubts about that force and power, you should remember that a certain unnamed President of The United States, with all his Executive Perogatives, tried mightily and unsuccessfully to get the FCC to eliminate the four licenses to broadcast of the Washington Post Company. He failed. Utterly and miserably. No presidential fiat, order, request or demand was sufficient against the great bulk of Old Granny Commission.

I have decided to visit WTOP-TV and assist them in the use of their precious, some would say all-too-precious, Channel Nine. If you ask the experts in the broadcast buy-and-sell business, they will tell you that that particular frequency, in that particular city, is probably worth some $35,000,000 to $45,000,000. That reflects on the effectiveness of the government to protect that monopoly. That reflects, in turn, on the net return on invested dollars which is so phenomenal that broadcasters—through their singular power over Congress and the FCC—have seen to it that the latter will, can, and shall never, never, *ever* reveal the sacred figures of profitability which are filed with that government body each year by each broadcaster. It's a good thing too—that those terrible figures are kept secret. If the public knew, if the public only *knew* how their air was being sold—they'd probably go into open and angry revolt.

Anyway, since I know that the air used by TV Channel Nine is really my air, it belongs to the people, dig, I take a bus up to the northern part of the District of Columbia—to drop in on the station, tell them I am in town, available to talk, or sing, or dance—whatever it is that they may want to help them use this frequency more effectively. Or, as the broadcasters have it *ad nauseum* in their filings with the FCC, to perpetuate their hold on our air, The Public Interest. I figure that the public will be supremely interested in what I have to say, so I look around for the entrance to Channel Nine.

It isn't very easy. In order to help us, you, me, the public, find our air—WTOP-TV has put themselves in a building not unlike, say, the Los Angeles County Courthouse and Jail. And to help us find our way, they have nothing outside the building to tell us the true nature of their business. No sign that says "WTOP-AM & TV," for example. Or "Channel Nine." Or "One of Your Government Granted Local Monopolies." Nothing like that to help me find my way to their door, to their lobby, to the "Receptionist." *Oy*, their receptionist.

The Channel Nine lobby was invented in about 1954, and remains unchanged to this very day. Brown and tan walls. Brown and tan! Yellow and pink pillars. Yellow and pink! Carpet the color of dog doo. Ragged, too. I think they put on the dog like this to let us know that running one of the top-rated television stations in one of the richest markets in the country ain't no picnic. And in case you don't get the picture from the walls, and the floors, and the pillars—they put Molly the Terrible Tugboat right there, at Reception, so you can get the picture.

Those of us who have been in community open-access radio for awhile tend to forget the Maginot Line that is set up around the entrance to most if not all of the commercial radio and television stations in this country. I thought of KCHU: you come in, and you wander around, and if you feel like it, you drift into the control room, and watch the on-the-air person doing his or her on-the-air stuff. If you feel like it, you can probably join in. If you feel like it, you can even do, as some unfriendly folks have done, threaten the air people with some mayhem if you don't like them, or the station. We pay the price—but the value of the station has to be measured by the degree of access.

To Molly The Mountain, "access" was one of those funny words. Like "antidisestablishmentarianism." Or "omphaloskepsis." Or "forensically pugnacious." She had never heard of—never dreamed of—access.

"Hello," I said—assuming my best Robert Redford manner, "I'd like to go on the air." It took a moment, a split-second, for her to shift the wad of gum from left lower jaw to right lower jaw, and move copious gams around on the lean-back white naugahide chair some. "Do you have an appointment?" said Molly of the hams and gums. "No," I said. "Who are you looking for?" she asked. I do believe, although I am not sure, because of the excitement of the moment, that her eyes darkened slightly as she spotted in my character the general ne'er-do-well elements that all my friends have come to know and to love.

"I just want to go on the air," I said. I could feel my copious gorge rising. "Who do you want to see?" "All right—let me see your Public Affairs Director." "You can talk to him on the telephone," she said—and she motioned to the yellow telephone on the plastic table in the middle of the lobby. She punched it up on her dial-a-board, and it rang stupidly while I looked at her, stupidly. "Go ahead, answer it." Who am I to deny the tolling of the bells. I picked it up—stupidly—and she dialed the Public Affairs Department. The Public Affairs Department was all in conference. As was (I know, I answered the 'phone each time she rang it for me) the Community Affairs Department, the News Department, the Programming Department.

"Look," I said, getting as close to her as I dared, "why don't you just let me go on in, and wander around, and see if I can find somebody. You know—just buzz the door over there (the door to the aether was and is and always shall be locked)—and let me go upstairs and talk with someone about my ideas."

There always comes a time—when you are talking to someone in power, or someone who has delegated power, or someone who thinks they have power—there is always that time when you come somewhat close to the edge. For all I knew, Molly of the Large Knees might have had a button (near those large knees) which was red, and big, and convenient to her wandering hands—a button that would bring the security people on the run. And dear God, you know me: I would be the last one to want to create a scene. If they came—and tried to take me away, I would be scared, wouldn't I?

Fortunately, Molly didn't go for the button. Not yet she didn't. She widened her eyes some at my suggestion, and told me about Policy. Between Molly and her fat wad of gum, she told me a great deal about Policy. "Policy doesn't permit . . ." "It's against policy . . ." "We have a policy . . ." Fact is, Molly told me so much about Policy that I figured it was some big mushroom, or marshmallow, or baboon, maybe—labelled *Policy*, and if you tried to do something wrong, this mushroom, or marshmallow, or baboon would roll all over you, and squash you, and imprint all over your body in purple ink: *"He tried to violate Policy!"*

All the while Molly and I were carrying on this discussion, this altercation, really—the TV set in the lobby was blatting away, behind me. Pictures flashing in washed out color; lots of sound. Something called "The Match Game." What happened was that some coarse-faced MC would hand out a mythic, classic problem to this bright-faced panel. And they would scratch their heads, and write down something, and hide it on their desk, and then mumble something, and show their sheet of paper, and the audience would go into uncontrollable laughter. "Sam The Super Salesman rode into the station on a *blank*," said Craggy Face not once but seven or nine times. The audience loves it. The panelists love it. Great gobs of laughter and crazy howls of glee. Everyone loves it and, judging by how often her eyes slip past gesticulating, talking me to the TV set behind me—Molly loved it. She certainly loved it far more than this dibbly weirdo standing in her puce lobby, taking up her time and trying to fuddle her thoughts with those dim concepts of access. Access your granny. Behind Molly, on a tiny table, under a tiny lamp, stood two tinsel Christmas angels, mouths a round O, filled with the Spirit of Christmas Long Past. Christmas in July—and I am sweating for all I'm worth . . .

I must have spent 45 minutes out on Wisconsin Avenue, sweating profusely, at 96 degrees in the shade, waiting for the #32 bus. Molly must have called DC Transit, told them there was some crazy out in front of the WTOP Building, that if they knew what they were doing, they would send all their busses back to the barn or garage or busarium—wherever it is they send empty busses in 96 degree heat in the middle of summer, in Washington, D.C. It gave me a good chance to think on American Television, and the American system of licensing, and what it must mean to some people who want so, who want *so badly* to get their ideas conveyed somewhere beyond the local street corner, the local bar. I had plenty chance to think on those television station owners—Kay Graham, & family—isn't that the name? This year, "Broadcasting Magazine" is beside itself with joy, describing the immense profitability of television stations all over the country. And those figures go treble for the 4 or 5 television stations that have the good fortune to be located in the top 10 or 15 markets in the country.

Me, and the sweat pouring down, thinking about Molly and her hams and tinsel angels. It ain't her fault. Somewhere, upstairs, at some time, maybe twenty years ago, when they first began to realize the fantastic publicly-granted mining operation they had on their hands—they decided that there would be a lock on the door, and a woman, a tough one, who would be the keeper of the gate. To keep out all the kooks, with their kooky ideas of communication and transmission. Those ideas, ideas that might be a bit more pertinent, more meaningful; telling us what the world is about, what we are doing in it, what we should do about it. More meaningful, some would say, than Sam Supersalesman, riding into the station, on his *blank*. There, me, in the supersun, that superfierce July day: me, Super-Unsalesman, just having rode into the station, with a new idea, that might have been worth something different. Me, come riding into the superstation, riding straight into SuperGam, who would not now, nor ever, let the unwashed, unknown, unannounced, into that great rich three hundred kilowatt *blank* which is Channel Nine, blatting into 4 million brains, in that all-too-important center of power, known as Washington D.C., center of most of our dreams and all too many of our impossible failures.

August 1976

THEY WORE A BIRD ON THEIR SHOULDERS

WOODROW WILSON campaigned to keep us out of Europe, out of The Terrible War, The First War. Franklin D. Roosevelt campaigned as a fiscal conservative (in the worst of the Depression) then as a fiscal liberal (in 1940). It was only shortly after that General Motors, Willys, Kaiser and U.S. Steel were given $800,000,000 to make us the arsenal of democracy—and it was only a while later that the heavy machinery bought with tax money was turned over to them at a penny on the dollar so that they could make their post-war fortunes.

Richard Nixon ran in 1960 as a hard-nose against the Chinese Communists, and twelve years later was having lunch with them, in Peking. LBJ won countless votes in 1964 by accusing Goldwater of being a war-monger—and look where *that* got us.

Adam Smith said that history is never kind enough to repeat itself—but I had to think of these matters as I watched the "debates." The joke is that they were called debates—but what we saw really were Kukla and Fran acting out their predetermined roles for us. Michael Arlen pointed out in a recent article in "The

New Yorker" that we are all too TV sophisticated to have left any surprises in personality revelation during such confrontations. Subject is now dominated by the reality of presence: the natural energy and diversity of men aching for power is crushed under the need for extreme care.

Or, to put it another way: we resent the plastic surface care smoothie appearance, publicity, speeches, of the Nixon men: yet is it not made such by the demand of that single fish-eye lens poked into their faces, a lens which transmits their vision to the hundred million, with microphones that accentuate every human flaw? The tragedy of Nixon was the tragedy of any man who manufactures a vision of himself and has a madness for secrecy to protect himself from revelation: when you try, and try hard, too hard, to be straight, secret, hidden—then the very media that found you will destroy you. In this case, the mechanical means to protect secrecy provided endless tapes to burst the symmetrical sphere so carefully built around him.

Those of us who watched National Public Television at least had the diversion of watching the fluttering wings of the translator for the deaf. With black sweater, the visibility of sound was perched on each of the shoulders of the candidates impartially, equally: and the flashing movements at least provided some entertainment, some diversion. We could hope that the next session would have an alternative visual channel, perhaps with a rock band, perhaps a few reminder shots of defoliated Viet Nam, maybe a picture of the money these two klutzes cost us, and will cost us, when they get into the business of spending four hundred fifty billion of our dollars.

And there we go, getting trenchant about a non-event. Not only were the debates non-events of the first water—the election itself can be considered such. Nixon always grumbled that he had control over no more than 15% of the federal budget—the remainder was pre-spent by previous administrations. He could, in addition, appoint two or three thousand people a year to governmental posts—but he knew that the 4,000,000 civil service employees were not going to be affected (or even listened to) by these temporary civilian "political" overseers. The miracle was not that Nixon did so much to trample our civil liberties at the IRS, and with the connivance of the CIA and the FBI: but rather that he had so much problem getting such things done. After all, he was The Prez.

If I didn't know any better, I would think that the presidential election process was some sort of poppycock—designed to keep the poor blind electorate from thinking of or knowing about the real deep dangerous divisive problems of the federal government. Like: why is it that our one-time hero Roosevelt, lover of Democracy, master of the four freedoms should create such a giant overblown vulgar disgusting fat klutzy poop-pie of a government so insensately out of our hands? Why is it that you and I ship $450,000,000,000 out to the Potomac each year—and get to see so little for our bucks?

Or: as an alternative valid question: if we plug so much money into this juggernaut called "Washington"—howcum we get so little for so much. Someone does. Get a bunch of favors, that is. We don't: at least we won't as long as they won't let us live in the United States of America (Delaware, Texas, Nevada, New York, Montana)—but rather force us to live in the United Monoliths of America (IBM, US Steel, General Motors, Lockheed, Skelly Oil).

Somewhere along the line they took away our chance and our hope of affecting the federal government. The public weal is still taught—I suppose—in the 9th Grade, but a fat lot of help that is to the rest of us. They throw us Carter and Ford as if they, or our vision of them, could really and truly change the frustrating beast that slobbers, drools, and snorts there somewhere between Virginia and Maryland—but anyone who reads any of the great muckraking journalism out of the 1970s knows that at best the machine of government consists of 535 fairly benign legislators being well-fed and well cared for by 5,000 lobbyists (outside the government), another 5,000 (from within the government—even the War Department covers Capitol Hill like the dew), and a couple hundred poor folk lobbyists from citizens' groups *who don't have a chance.*

The Free Enterprise Institute wrings its hands and worries that 73% of the people think that business is out to screw them, that the government is of the people, by the people, and for the people that run the giant multinational corporations. The people think right: so many of us have been brutalized by advertising: a business at work to crush our credibility and test our credulity. Now the FEI is going to use advertising to convince us of the worth of Free Enterprise. A fitting companionship.

Most of us sincerely believe in Free Enterprise. We value what it represents for the individual. Hell, most of us would give our eyeteeth to live in a country that honored and believed in Free Enterprise. But when we are faced with the wrenching knowledge that United Fruit runs the CIA, ABC runs the FCC, Lockheed runs the State Department, Mobil Oil runs the Texas branch of the legislature, and the telephone company runs them all: when we are faced with that bitter knowledge—we are forced to the droll conclusion that we could only have free enterprise if the big corporations would get the hell out of the way. As long as they write the laws they work under, frame the taxation rulings that they are to follow, infest the regulatory commissions that are supposed to regulate them: as long as they do this so richly and so well—the rest of us won't even have a crack at "free" enterprise. It's just too damn expensive.

We watch the two mechanical men act out their show on television. They say that some 100,000,000 of us watched the performance, and were well-bored by it. It is a necessary adjunct to the playlet acted out every day there beside the Potomac. Blithering about free enterprise while the poor and the middle class are taxed without surcease for the benefit of the Carters and Fords and those miserable jokers who give us daily lectures on the wonder of our system of government. Thank god for the birds perched on their shoulders: the beautiful blackbreasted white-winged birds that fluttered and played on their shoulders while they, what they call the "candidates," carried on about nothing particular at all.

November 1976

TOURING THE WORLD'S LARGEST RADIO STATION AT PORT JEFFERSON, LONG ISLAND, 1921.

AN ELEPHANT BROADCASTING, 1925.

KPOO

TALKING to Meyer Gottesman was like bobbing for apples. There'd be this vat, with all these ideas floating around. Sometimes you could sink your teeth into one, but more than likely, you'd get this face full of cold water. "What the hell is he talking about?" you'd wonder. Meyer talked nonstop.

When I first heard about him, he had just made application to the Federal Communications Commission to set up a radio station in San Francisco. I invited him down for an interview, and tried to tell him he didn't have a chance: that there were no holes in the aether for another FM station. Good thing he talked nonstop and didn't have a chance to hear me: I was wrong.

I figured it had something to do with his corporate name. Meyer put an ad in the Berkeley Barb, asking for a free lawyer to help him put together a radio station. Sherman Ellison—one of those dope lawyers with Rohan & Stepanian—called him up. That was the way Meyer got things done: he asked for it. Directly. Nonstop.

When Ellison went to work to put together the non-profit non-stop corporation, he asked Meyer what to call it. Meyer said: "Poor Peoples Radio. That's who it's for," he said: "poor people. So we'll call it Poor Peoples Radio." Simple. Brilliant. I am going to guess that that is why the operation was so successful, at least at first. In getting the tax exempt status, both from the state and the Federal IRS. In getting an organization together. And in getting a permit from the FCC to build a station.

Meyer used to believe in asking for what he wanted. Anywhere. He had that child-faith that believes if you want, and want bad enough, and ask long enough—then you'll get it. After he got his contruction permit, he went out and got another permit . . . from the San Francisco police: a begging permit. Meyer wanted to take Poor Peoples Radio to the People. He set himself up on the street with a tin cup and a sign around his neck GIVE TO POOR PEOPLES RADIO. That's what most people remember about that early stage of the last radio station to be squeezed into this (or any similar-sized) market: a funny man with dandruff and glasses, the thick kind so the eyes are huge like they were behind magnifying glasses, and the sign around his neck and the tin cup. Which he rattled in a professional manner. Ask, children: and they will give.

Everything fell apart after that. Those of us who are into putting community radio stations on the air recognize what we call Phase 3C . . . The Construction Doldrums. You get the permit from the Man, and you get through with the fun part which is calling up everyone who ever doubted you and saying "We got our C.P." and then what do you do? Because building that dream station is damn near impossible unless you are rich. Especially for poor dreamer types like Meyer. It's one thing to go out on Broadway and rattle a battered cup and take in pennies. It's another thing to buy a transmitter, and stick it up on the Fox Plaza Building, and get a four-bay Jampro antenna, and run a remote control unit down to some would-be studio, especially when you don't have the damn $175 to rent that would-be studio.

So you think: what I should do is to get a Foundation to fund this operation. Foundations are crazy about Poor People, aren't they? That's what Meyer thought.

Meyer was a child of the aether and didn't know that Foundations—from the Ford Foundation down to some dinky organization like the Cynthia G. Motherluck Memorial Fund—are set up by people who want to give money to their favorite projects. And they look at all those 173 submissions-prospectuses-requests-demands that come in the mail every week as a big pain in the ass. Those foundations no matter what they tell the IRS and "The Foundation Journal" are primarily in the business of making grants to their friends on other foundations.

This contributes considerably to the perpetuation of the American Dream-Delusion suffered (painfully) by people like Meyer Gottesman who figured that those rich San Francisco Foundations would love to fund a radio station for the people. For the people: all those who had no voice in the aether-babbling buy-sell-push-me-pull-you known as American radio. Foundations would gladly fund radio for the people. They would. Wouldn't they?

My ass. And it was this single reality that drove Meyer from Poor Peoples Radio, Inc., and almost strangled the whole thing before birth. Reality. Money reality. The cruel buck. And the terrible realization that no one really gives a damn about another radio station—the last radio station—in the nation's sixth market.

That was when the KRAB people got involved. KRAB is a Pacifica-type station in Seattle which bears a major difference to the KPFAs and KPFKs and WBAIs of the country. That is—it is apolitical enough that it can help to foment new non-institutional community non-commercial free-form radio stations around the country. It doesn't have the brand, given rightly or wrongly, to Pacifica by the FCC. (Pacifica has been trying for five years to open an FM station in Washington D.C. In an appalling travesty of justice—although Pacifica is in every way legally, technically and financially qualified to build such a station—they have been blocked by the FCC's Broadcast Bureau. For five years. That's what happens when you get a 'name' in radio—a name for trouble-making, shit-kicking radio.)

The KRAB network has set about establishing community stations all over the country—and has succeeded in Seattle, Portland, St. Louis, Pittsburgh, Santa Cruz, Los Gatos and Atlanta. We had the expertise and the ability—and could borrow the funds to get the station on the air.

Meyer turned the construction permit over to us and gave up. The last time I heard from him, he was working as Chief Engineer for one of those syrup top 40 classical-warhorse stations of the KKHI school.

Our job then (this was back in 1971) was to get the radio station on the air as quickly and as cheaply as possible. We wanted a station that would have the power to be heard in all of San Francisco—and would have freedom from the political babble-rabble school of radio, as well as freedom from the educate-'em dead school of institutional broadcasting.

With these simple goals, and the fact that we had put some eight stations on the air in this country over the last 10 years, you'd think that this one would have been a cinch. After all: think of the enlightened population of San Francisco, where freedom of speech is of epidemic proportions. With all that kultur and those intellectual smarts, popping out like pimples on some teen-age face, it should have been a lark. Lark, my foot.

I won't even begin to start to think about telling you about the agony of the last three years of our discontent with putting this thing on the air. Not only does the story now bore me (I resigned three times myself during various directorial arguments) but I refuse to open myself for libel. That's how hot our passions ran. We knew it was the last station. We knew it had to be good. We knew that this was the last chance for San Francisco radio—no matter how tiny. We knew we shouldn't let this operation fall into the hands of dingbats that abound here. In such profusion.

I think I'll lay a good 50% of the blame on the city itself. You know what San Francisco is like. When we're not busy leaving our hearts on Nob Hill, we are busy murdering the freedom and spirit and excitement of whatever it is we are after. It has something to do with the wind and the fog and the inversion of the season: it turns our minds to chicken noodle soup.

The energy in this city is on a par with some 53 megavolt holding substation for PG&E. You can hardly say "Boo" (or "Poo"—we named the station K-POO; as in "Pooh, Winnie ther") without all sorts of people calling a meeting and discussing your statement until three in the morning with 12 floor votes and two enabling orders.

With all the crackpots and political dildoes this community nurtures like mold, it's hard enough to get across the street without being speechified to death: much less set up a free-access radio station. I'm not going to tell you about the letter about KPOO that went out to the FCC and the IRS and the ICC and the ICU and the International Order of Moose. I'm not going to bore you for Christ's sake (like we were) with those endless dickering babbling meetings with people shouting at each other and getting worn out like old shoes.

KPOO went on the air June 6. Like I said, it's the last station that will ever go on the air in the Bay Area unless someone invents an electromagnetic shoehorn. It's stuffed down at Pier 46A but in keeping with the tradition, spirit and history of Poor People's Radio, it has been handed an eviction notice by the San Francisco Port Authority a bare week after going on the air.

Where it is going to go and how it is going to do it is beyond me, but three years with KPOO has convinced me that it is like some drunken uncle: you wonder how the old geezer lives from day to day what with the way he runs into walls and telephone poles and retches up his guts twice a night. But he survives and so does KPOO and someone is going to call us up (like you) and say that there is a free room down on Grant Avenue where we can perch an orphaned and definitely unlovely community radio station for a year or so.

Programming. That's what we should be talking about. What is KPOO going to do that's different than KPFA or KQED or KSAN?

Bless me if I know. The important thing is access. Whether you know it or not, that damned KQED-FM with their 100 kilowatt transmitter spends so damn much money on flabby bureaucracy and deadly dull programming that they will charge you or me $35 an hour to go on the air. Which is a travesty of the whole educational radio concept: radio stations have a positive duty to open their microphones to the world out there, beating at the door. And for an educational station to charge for time is as evil as those million-dollar stations like KSFO and KGO and KCBS to demand money for their time.

So KPOO is and will always be a free-access station. To those people who have something to say, and no place or way to say it. The station is committed to giving time to the public, poor or no. Everyone gets 15 minutes. If you are good, then you get more.

And I don't mean time to go on the air and play Joni Mitchell or The Silver Strings or Stan Kenton or Grand Funk. The station will and should never compete with the existing stations. Its job is to supplement them. (The idea is not new with us: when Lew Hill set up KPFA in 1949, he saw it as a broadcast service that would give air to all the unknown and unpopular words and musics ignored by the commercial and 'educational' stations. This still may be true of the talk programming: but the music of KPFA has fallen into the pasty hands of George Cleve and his gang—whose idea of an original work of art is another presentation of Brahms' 4th Symphony. For the 400th time.)

For that reason, when KPOO plays records, it will strive to air music from all over the world ignored by existing stations. Music from Sunda and Serbo-Croatia and Chad and Venezuela. Music which because of its very differentness is left unplayed by almost all radio stations in this United States.

But the music programming is and has to be secondary. What KPOO will be (and is) specializing in is what they used to call—on commercial radio—Local Live Programming. Street Fairs. Interviews. Talks. Live presentations of the San Francisco Board of Supervisors meetings (to go on immediately—or as soon as you send us a check for the telephone line charges).

Talk. Great gobs of talk. Not that preprocessed talk, that Wisconsin Cheese called "talk-radio" where your comments are guarded, husbanded from anything too shocking or different by a 'producer.' No: I am speaking of robust, free, diverse, controversial, wide-open, free-form discussion and questioning and answering and thinking and verbalizing and wondering. All over the air, live, and fresh, and unafraid. The words of an angry confused populace. Who—up to now—haven't had the free access to 100,000 or 1,000,000 radio sets. There were always a few hedges on the chance to speak freely in radio in San Francisco. Always something in the way. Like money to be made, or a producer, or a director, or a board of directors.

No more. KPOO lives, and it lives to feed a million free words into the willing ears out there. Free speech: the ideal of a free republic, so often obfuscated by fear and trembling. No more. Pier 46A. Just south of the Embarcadero. The little shack behind what used to be called the Deep Six Restaurant. Deep Six Radio: KPOO. Full fathom five.

Why are you just sitting there? Why aren't you helping us? With your station?

The Bay Guardian, 1973

KCHU

WHEN Dennis Gross first came back to Dallas with this hare-brained scheme for a radio station, his mother is reported to have listened to his plans, thought for a moment, and said "So? What are you going to eat?" Here, some four years later, we are still wondering.

He did remind her, however, that when he was a student at Greenhill School, he took over the school poetry magazine *The Pimple* which had run in the red since its inception. Within a year, it was in the black. At least until they struck a gold medal for him for making it successful.

Dennis had worked at KDNA in St. Louis for a year or so. That's where he got the idea for KCHU. KDNA was a ratty, hole-in-the-wall community-access station which operated out of a brick fortress stuck in the middle of the black ghetto. The station had, through its grumpy programming, managed to irritate most of the St. Louis establishment.

KDNA was pieced together and run by a misanthropic 11th grade drop-out by the name of Jeremy Lansman. He couldn't spell "open-access" or "free-forum" or "community"—but he managed to create a radio station which made these a reality for St. Louis, In its brief 5-year life, KDNA was probably closer to the ideal of what American radio could have and should have been: what it might have been before all the commercial turkeys came along and sold it down the river.

KDNA was fearless and magical. It returned the community to the radio studio; it opened its doors to the poets, and musicians, and political freaks—the dispossessed and longing who had been in their seedy rooms too long, waiting for someone to come along and spring them from their isolation.

KDNA was radio reversing itself: asking that the people who lived in a city bring the city into the radio station, and cascade it out to the far reaches of men's minds and the horizon. A compelling idea. One so old (and yet so forgotten) that it is new. At its best, KDNA could count on 200 people a week and volunteered around the hard-core of a dozen or so who made the station work. They created the words and ideas and feelings that came out of the hot-house on Olive Street. We could hardly believe our ears.

That very persuasive concept of *community* grew a spiral out of KDNA and came to create a dozen or so others—including stations in Columbia (Mo.), Pittsburgh, Madison, Atlanta, Columbus (Ohio), and now Dallas.

Community radio, to those of us so deep in it, means that there is (or should be) a hole in the dial for us to stuff our words or musics into. Most of the holes have gone: into the hands of the *moneymeisters* who sell our ears at cost-per-thousand to those who want to sell their wares to us. There are others which have gone to the religious folk, who don't seem to cotton to voices or experimentation or diversity, or worst of all, to the professional educators, who are as wily as all the others in keeping the folk out of their studios, the talented amateurs off "their" air.

That's where KDNA and WYEP and KOPN and WRFG and WORT and KCHU come in. Stations that need, depend on, want, the talented, untaught, cheerful, wide-open, free, uncorrupted, rough-around-the-edges volunteers. The new industrial state called America has created an economy in which very few are actually starving, in which very many very talented people have many many hours days weeks months on their hands which they are willing to invest in non-profit, alive, thought-moving, mind-expanding pursuits. Like radio. For those of us who are the beneficiaries of a whole new economic (but unmonied) leisure class—it is the experience of direct communication with our villages of the mind, and it is called "community-access free-forum radio."

Gross left KDNA and St. Louis in August of 1971—and spent the next four years of his life going to form school. Not reform school: form school.

Let's see: there's the application for construction permit form from the Federal Communications Commission, and the application for STL and SCA form from the same body; there's the Federal Aviation Authority form to construct a tower. The Internal Revenue Service has an army of them for tax-exempt status application, as does the state of Texas. Cedar Hill had a few for the transmitter site, and the city of Dallas had a few more for the building and for solicitation of funds. You know what I mean by Form School now? We think there is one we might have missed, from the Great Aether God, but since we haven't gotten the call, we are pretending it doesn't exist.

If you took all the forms, and stacked them all together, and took Dennis Gross, and stacked him next to them, they'd both stand at about 4'9" (he's been somewhat shy on height since the operation) and all of them seem to represent nothing much more than sheer, animal *will*. That means that one person—at first—and then a group (that's the board of Agape Broadcasting Foundation) and a whole bunch of friends and then all the people who come in later because they are interested in radio; this type of radio: what it means is that all these people are willing to take the time and the thought to fill out the holes in all these blank pieces of paper, fill them out with words and numbers, and file them here and there because that represents something to that someone who has to say either 'yes' or 'no'. And you have to stack up a whole bunch of them and send them downtown or to Austin or to the biggest paper mill of them all, Washington D.C. And when finally there are enough words and papers and filing numbers someday, when you aren't even really watching, really, someone pops up and says: O.K. You're on. And you are.

Someday, we'll take all that paper and make it up into the history of KCHU (even though the history grows by itself, now that we are here). We'll add up all those pieces of paper and make something of it. One of the things that Dennis Gross remembers from these four years is himself always telling people "We'll be on the air in six months." He said that for four years. It was easier than explaining about all those hearings and filings and stuff.

But there is a grail at the end of the line. I mean, if you are in the government-filing business because you are working on a monopoly or a $4,000,000 contract or an oil well—the grail is profit or capital gains. The grail is a bit more obscure in "community non-profit radio". We know what it is, we who knew it in Seattle or St. Louis or Los Gatos. But we didn't know it in Dallas, and it was hard as hell to explain to people what great, diverting, diverse, deranged, dynamic, dingbatty radio is all about where there is no model. I mean you can say it's sort of like KERA or KRLD at night or WRR-FM—but you can't *really* say. So you don't. And they say (one more time) "when are you going to be on the air?" And you are fixing tape recorders to pay the goddamn telephone bill and you say, "O, six months." Sometimes you are not really too sure. I mean when that Christly Channel Eight and A. Earl Cullum dump a 25 page PETITION TO DENY on you via the FCC—then, maybe, you think, it might take more than six months. Now why in the hell did they do that? What's wrong with them. Fuddling up our dreams.

But then, maybe it was Fall of 1974, things began to swirl together. Out of the fog something's beginning to coalesce into a ghost-whirly shape which some optimists might call A Radio Station. After we indebt ourselves another $150,000 (land, tower, debt into perpetuity) those nitwits drop their creepy petition. Cedar Hill zoning board tries to put us back another year but then the town council oks a building permit. FAA says they won't protest a 700 foot tower. Things begin to move together. All that paper means something, doesn't it?

The property on Maple Avenue that you have been eyeing for a year or so, the old Haunted House, comes down in price. They accept $5000 down on it—and another $100,000 in indebtedness. Collins makes up a do-it-yourself equipment kit for you (start payments in October) and you sign your name and you are part of the Great American Dream: all aspiration, no assets, illimitable indebtedness—enough to tie your children and their children into eternal penury forever. Amen.

Remember last mid-winter the day after you moved into the building on Maple and the pipe burst on the third floor and you think "I'm going to Peru." You don't—but all that water cascading niagra knee deep onto the first floor is enough to make you weep.

But I guess it was March or April when the whirl coalesced definitely into the shape of a radio station. Bobby and James took almost no money and somehow made a couple of studios on the second floor. Cynthia made the books make sense. Larry started sticking pieces of equipment together. PiRod Tower Company sends a half-dozen monkeys down from Ohio and they spin this piece of tower together and then we go out to the site and tie on a big one with a keg of Coors and remember their stories about the windy reaches of Wisconsin from 1500 feet up.

All of a sudden the transmitter comes in the door, and the transmission line. Two or three volunteers come to hang out at the station and work (we call it "the station" now. "The station." "The station." We try it out a couple of times to be sure it works, like winding up a new clock). They not only come in—they come back: a core group of station people begins to coalesce. I tell you—there is some magic in it. We are working on the magic aether (which may not exist)—and part of the magic is those who hear about it, who are ready for The Concept, who have been waiting for it, really. Who come down to The Station, and come back. We are all in this magic-pot together. Alan paints up a sign that we hang in the front yard "Community Radio/KCHU/90.9." We have a label. We are here.

Magic: one day Larry is out at the transmitter working. You know Larry's secrets. He never tells *anyone,* the creep. But someone wormed it out of him. That he was going to push the button. So we went home and sat around the old KLH (just like us kids cross-legged in front of the Philco, waiting for the word). Sat around, waiting, Waiting waiting waiting. Four years waiting. Waiting waiting.

We tune the radio to where we should be. Listen to white noise. God, we've been listening to white noise at 90.9 for years.

Not just yesterday or last year or in 1971 when Dennis came up with this ridiculous idea. No that white noise has been there . . . forever!

It was there before we were here. That white noise was there, at 90.9, just puttering away, long before we were born. It was there when that crazy Edw. Armstrong first started trying to displace the white noise in Alpine Hills, New Jersey in 1933. It was there when that zany Reginald A. Fessenden started 65 years ago puttering with the aether in Brant Rock, Mass., or when Stubblefield made the land talk to him in Kentucky in 1894.

That white noise was there (just whispering away) when those tobacco-chewing cow men first came pothering into the Trinity River area looking for some acres to plant; it was there, going sssssshhhhhh—and they didn't even know, didn't even dream it was there. It was there when seven million bison ranged over the area, being born, living, dying white bones with only a few human figures seen barely off in the distance.

The sssssshhhhhhhh of the universe (the hush of the whole nights of eternity) was there before we were here, before the first bloody man was ever conceived out of the great grey-green-greasy goop that made us all: that whisper was there even before the hot stellar rocks came whirling together to give us an Up and Down. That whispering of the universe was there before . . . before . . . before (maybe) even before then. And that nutty Larry comes along and with a sniff and a push of some crazy button banishes that white noise. Just like that: we're sitting on the floor before the old KLH listening to the whispering of the ages, and

Larry sniffs and pushes the button and there's a surge and the power transformers hum a little harder and

zip . . .

All the whispering is gone. Good Lord! And there's Larry (who had the foresight to take a microphone out to the transmitter site and plug it into the transmitter) mumbling about wheatstone bridges, and Mhos and Ohms, and modulated universes, and he says "K C H U, Dallas." And, of a sudden, that hole in the dial has a name. And it's us. Us.

Those who should know say that the birth of KCHU has been easier than almost all the others. Before, in other cities, they either didn't know what they were doing, or they screwed up their organizations right from the start, or they got a dose of broadcast clap (interior politics) from the beginning, or there was a breech birth, or something. KCHU had a board of directors that *cared*, but in their love didn't try to strangle the baby. The core group worked together, with a minimum of friction. It might have been Dallas—either the city folk were so eager, or so indifferent—that the naissance was like falling off a log.

Maybe there is something to be said for a would-be broadcast station going on the air with $350,000 in debts, all new equipment, all credit, the businessman's dream . . . It is a cinch that there wasn't a cent of federal funding that helped KCHU: Health, Education and Welfare and Corporation for Public Broadcasting were elaborately disinterested in a free voice for the community—even though our applications to them were as good as any of the others we put together. They don't much cotton to the disorganized unestablishment anyway—no matter what the depth of our commitment and interracial standing. If Thomas Jefferson and Patrick Henry and Alexander Hamilton had been around to band together to put together some sort of Voice of Dissent, it's a lead-pipe cinch that HEW and CPB would turn them down flat because of their funny ideas about the king.

So it's the debt load that makes KCHU different. For the first time there was enough confidence in the concept of community-access radio that we could go seriously into debt to finance it. This means that there are enough such stations around that the equipment manufacturers and banks don't think they are dealing with some sort of fly-by-night and are willing to accept paper on it. That's new: and some time in the next 12-18 months we will be able to know if it was brilliant or stupid to pay for KCHU out of the future. *Da mihi castitatem et continentiam, sed noli modo.*

So what do we have to show for four years and all those forms and that fabulous debt structure? Well, we have a new broadcast facility which has superb equipment, virgin equipment waiting to be tested and used.

It is a station which will try our ears. One which will revel in open access. One which will utilize any army of unpaid friend volunteers to keep it alive, and ticking, and *there.*

The station will revive the art of early radio which was known as Local and Live. One which will (and already does) fling open the doors of the aether to anyone, *anyone* who has something interesting to say, or experience, or communicate, or sing.

KCHU: from now on 90.9 becomes a center for informality and reason and thought and ideas and ideals which so far, because of artifice, or fear, or greed, or pettishness—have eluded the air, and thus our ears. Here we have a center of experimenting with the feel and touch and dynamics of a city alive, alive to itself, a feed-back system which constantly reports itself to itself, through tapes and live talk and telephone beaming in from all sides: the voice of a city coming to feel itself through its own pulse, the pulse of live people telling us about ourselves. People talking to people. Not talking at them, or down to them, or over them: but to them. People talking to people as if we all were (we are) humans who deserve respect and hope and the freedom to give utterance to our artistry, our needs, our excitement.

It will take awhile. I mean: for the first few months there will be too many records, too much recorded music, too many tape recordings from other stations. It has to take some time for the people to get used to the idea that the walls are down, and that the microphone sits here open as the sun, ready to be talked to: that there are all those minds out there waiting to be handed some free thoughts, and some feelings, and a touch of live, lively music. The habits inculcated in all of us by the restricted radio patterns of the past make it hard for all of us to adjust to this brilliant new-old concept of a spot of dew on the rose, the dial, that is there for all of us to come to, to admire, to be part of, to use, as it should be used.

No, slowly, they'll figure out; I mean—slowly *you'll* figure out. That we aren't here to talk at you. Or by you. Or to you. No: you're here. To talk to all of us. Excelsior!

KCHU wasn't built by one person—or even a dozen. It was built by a hundred persons—building on the experience of the community radio stations that had gone before it.

And it won't survive through the efforts of one or a dozen. It will take hundreds of people, with their ideas and experiences and words and musics—to carry it into the next stage of growth.

KCHU is built on faith. Faith in the worth of its concept of radio—and the faith that the city of Dallas will support it. The ideal of KCHU is so old that it is brand new: that the community comes to KCHU, and becomes the station, which in turn goes out into the community. It is a new and a worthy idea, and Lord knows that the city (and we, too) need it.

Cui dono lepidum novum libellum
Arido modo pumice expolitum?
—Cato The Elder

September 1975

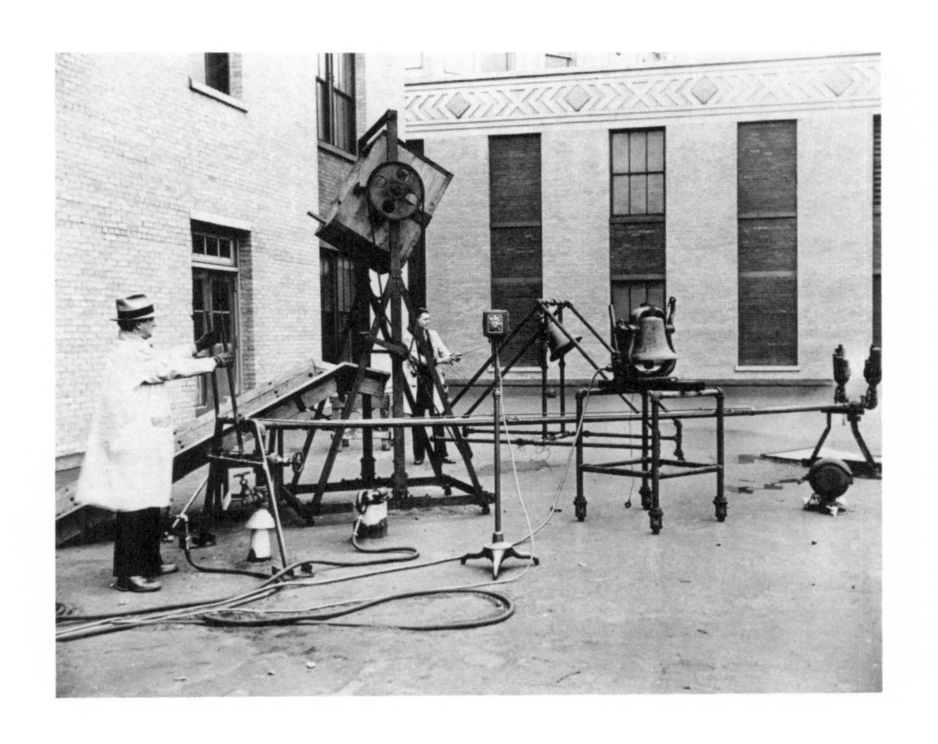

CREATING LOCOMOTIVE SOUND EFFECTS FOR NBC RADIO, 1930.

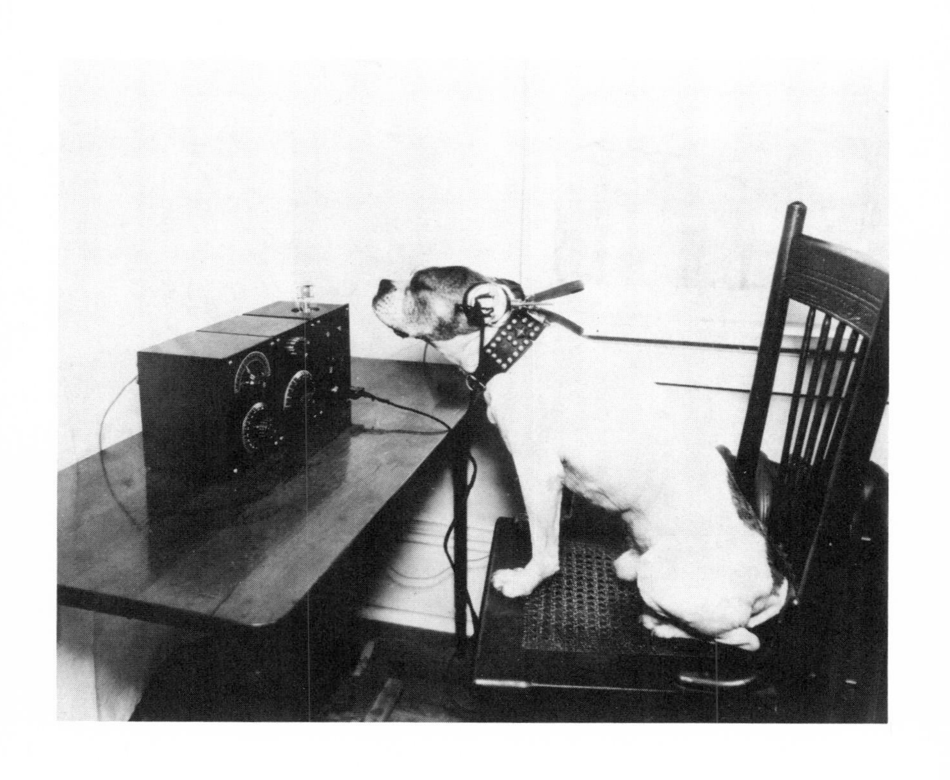

A DOG LISTENING TO THE RADIO, 1922.

LETTERS TO KCHU

Dear KCHU:

Last night I received your station from 7:55 PM (CDT) to 8:13 PM (CDT). I was listening to you on my Garamond Balloon Sporadic E Skit Rotograveur Receiver with a 41J Bullfinch antenna with pecker and isosceles triangulational with Whip-Bak Jelly Receptional Creme and Skin Tight Pleasure Dome Speakers with fingersnaps. The DeeJay was talking about Existentialism and Water Safety, and said something like the following:

"If you had blather the coming and the blather knowing but the place there is another blather which had known the other skather snakes and there was another one who tole me the one about the Indian with the snakeskreetch inside the vulpate when she announced her demise . . ."

I would very much appreciate your sending me a DX card with any attachments and gee-gaws that come along with your long distance listeners.

Belovedly,
R.A. Fastener
Cedar Hills, Tx

My Dear Sirs:

I read your program guide from afar and I note that your editor is fond of quoting Curzio Malaparte.

I should like to remind you that it was Malaparte who said: "Like all dictators, Hitler is merely a woman, and dictatorship is the highest form of jealousy."

When asked by a listener whether Caesar was, then, only a woman, Malaparte is reported to have replied: "He was worse than a woman. Ceasar was no gentleman."

Pleasantly Yours,
(Ms.) Cese McGowan
Monterey, California

Dear Mr. Milam:

I read with interest the materials associated with your letter of September 22, on the subject of broadcast license renewals.

Unfortunately I am no longer able to vote my views on the subject since—as you may have suspected—I resigned from the Commission last summer in order to become a bride.

However as Roscoe was telling me before I left "Charlotte, that Milam is certainly a boy with a sense of humor." We always thought that you were a good boy although you do get in trouble now and then and get those fellows over at the National Association of Broadcasters angry with you. My how they carry on!

Well, I am going back to the "Breakfast Club" (with an audience of one) now. I hope you appreciated my performance. I voted right along with the boys, and Dick said to me when I left "Charlotte, how did Don McNeill ever give you up?"

Respectfully Submitted
Charlotte Reid
(Formally of) Washington, D.C.

KCHU:

It came to me the other day as I was listening to one of your announcers that you really are in my radio. When I was little, we used to talk about the little people that lived in the radio—and for the music, there was an orchestra of Super Midgets. They had tiny violins, no more than an inch across; and these microscopic flutes, no more than the width of a hair, and their tiny tubas, that made a tiny croaking sound. Their kettle drums would be about as big across as the tip of your little finger.

I was alone in my room the other day, and was listening to you on the radio, and you were playing some music from the Congo, music for the (I think you said) Oogamoogui Tribe. I thought to myself: "That little band of musicians. They are no more than a couple of fingers high. They are dusty and parched from staying out in the sun, with their water buffalos all day. They have just finished crossing the veldt, and they are hot and sweaty; but they have decided to pull out their instruments and play some of their music. They pull out their M' Thumbri drums, and their gourd flutes. And they start playing—right there in the back of my receiver." They have lived there a long time—ever since I have had it, and they have just been waiting for KCHU to come along and get them activated. They took their instruments out of the hot sun, down to the front of the receiver, and they started banging and whistling away. This is probably their one chance of a lifetime. The rest of the time, they stay there, back in the brush near the I.F. stage—not noticed by me, or anyone else.

I am pleased that you gave them a chance to come out of the dark. No one else on the radio band had ever paid any attention to them. Thank you so much for your kindness to them—and to me.

Geo. G. Geiger
Plano

Dear Friends:

Last summer my boy Peter turned sixteen and turned away from The Lord. Please prey for His Salvation.

We gave him our love our bread and home. We preyed over him each morn and each night in that he could Stay Right With The Lord.

He was the nicest boy you'd ever want to see, but then he got to the TV and to them at the local school teach him Bad. They teach him to do things out behind the school and Lord Knows when I was at the school they never let us do That. And what they see on the TV with the killing, mameing, beating, whipping, mawling, beatings. I dont see how a child can grow up normal in this Sick Society.

Anyway starting last summer when he turned into a man he has gone wrong. He won't go to the church even on Sundays the Lords Day. I come to his bedroom and say Get Up Sinner! and he lies there and growns like he is haveing a spasm and his eyes bug out and I get scared go to his Dad and make him come get him out of the bed. He flails him with his fists won't act right and I am wailing 'What have I done wrong, O God? What did we do to deserve this fighting and fewding?' He and his Dad go at it like they was to kill each other and I have to pull them apart.

I know you talk on the air about youth Going Astrey. You must do something. Last Thursday that boy set my Bible down in an unspeakable place, and I rose up rathful and said "What you did you do Sinner?" He only cakles like a hen trying to goad me to kill him or worse. I swear to you that boy is going to get it with the Dyeing if he doesn't straiten himself out for us that Love Him in spite of His Sins.

Help us O Lord,
(Mrs.) B.W. Wrigley
Sulfur Springs

MODERN MAUD MULLER HAYMAKING TO RADIO, 1923.

MILKING WHILE LISTENING TO THE RADIO, 1923.

KLUK: SAN DIEGO, CALIFORNIA

KCHU stayed on the air for exactly two years: from 1 September 1975 to 1 September 1977. It was quite a good station. It played great gobs of original talk; we interviewed almost anyone who came in the door. We played enough ethnic music to raise a dozen ethnic consciousnesses. We programmed the first regular gay show in the Dallas-Ft. Worth area.

We were set up in a fine old building on Maple Avenue—which I bought for 5% down and the balance over thirty years—my favorite method of doing business. Because a factotum of the local "public" broadcaster had opposed us through petition to the FCC, we were required to spend megabucks to get a tower and transmitting set-up and land out at Cedar Hill outside of the city. I didn't worry. "When Dallas and Ft. Worth realize what a great new resource they have, they'll support us faithfully," I thought. Besides, we had applied for a $250,000 grant at HEW and—if we hung on long enough—that was a shoo-in.

The grant was actually ready to be disbursed to us in

January of 1978—but unfortunately, we existed no longer, so the money went elsewhere.

Dallas is such a hard place, and the hardness is so subtle. It isn't hard like a diamond, or a mountain, but more like a cyst. (One of the favorite songs played over KCHU was "The Oozing Cyst Blues," and it was quite appropriate.) You poke at it, and it doesn't yield, and you wonder what's wrong, what's *really* wrong. All the while I was living in Dallas, pouring money into KCHU's survival, I would wonder about the weirdness of it (and of me). And then I would wake up the next day, and keep on keeping on, being swept back ever further by the tides but, because I had no landmarks, not being able to recognize the approaches to the maelstrom.

People would ask me to compare it to the other places I had lived: Washington, D.C., The Bay Area, Seattle. "It's a very interesting city," I would tell them. "Not half as bad as you would think. Not half as bad as everyone leads me to believe . . ." Oh, Oh!

Dallas is such a hard place. Like a tumor. The minorities, the white society-changers, the black and chicano community organizers are taught to be tough and devious. They have to be. There is little reform or justice shown there, in the courts, in the streets. A member of a brown or black consciousness-raising organization is on the run for his or her life. There are many ways of maiming the human mind (harassment, thuggery, subtle violations of rights) and one has to be violently aware.

KCHU: the wide-open door innocent child. We invited all those folks in the door to proselytize. "We are the voice of freedom—this is what democracy is all about . . ." we said in our program guide, over the air. What we didn't see was that these people had been sledgehammered for just too long, in the name of Democracy. And to them, we were a convenient door-mat. The station turned into Serbo-Croatia vs. Bosnia, and both vs. Montenegro. And there was no Tito in sight.

Dallas is a hard city, like nails. Hate from outside came to be reflected in microcosm in our wide-open, free-form organization. We always like to think of ourselves as friendly anarchists, but street tactics came to infest the halls of that beautiful broadcast castle on Maple Avenue, and the station just wasn't old enough and strong enough to absorb it. And the people in the city didn't give a good goddamn.

I would wake in my house behind the station, and hear the mockingbird, thinking its song so beautiful—not knowing that I was listening to clever imitations of other songsters, and the bird I was listening to was a wind-up imitation brought before the court to amaze and amuse. And when I would drive through the streets of Dallas, I would admire the new buildings, which are uniformly built of mirror-glass. While you are looking at one building, you really see—distorted, twisted, tinted—some *other* structure. You try to look into the heart of the city and all you see are shadows, devious images of other forms.

We who have grown up in the media and have seen the spread of uniformity in this country come to believe our own myths. We think, for instance, that the country has been brought to some sort of a benign, common level by radio and television. We think that the noxious cities are coming to be more progressive—and because of that thinking, we ignore the rock-hard force of capital accumulation and the personality of communities welded together long ago by people who, for their own psychopathological reasons, were cruel, or stupid, or greedy. Dallas was shaped out of hot dusty humid flatland by people who had to be hard, who killed for entertainment, who screwed the poor and the natives for their own wealth. Their heirs, both spiritual and real, still run the city. They sit in their endlessly reflecting air-conditioned buildings and think tough as any cowpoke in a Gene Autry movie.

There is no blame. There never is. There's just success or failure. KCHU disappeared in a whirl of gun-toting anger, just like in the movies. *Qui custodies custodiet?* Who is going to care for those who are supposed to be out there caring for the rest of us? As the station was beginning to fall apart, I called in some people from ACORN. They said they would hold the station together, and run it. One of the station programmers, in tandem with the chief engineer, made a few calls to our major creditors to let them know what was happening. Collins (we owed them some quarter of a million dollars) became alarmed, and the ACORN solution came to be moot (big creditors carry big sticks). The station went "dark," as they say so scenically in the industry. It was the first community station to go off the air since KPFA disappeared for fourteen mounths in 1950. *Their* listeners banded together to get it back on the air; *our* listeners went out and got another six-pack of Lone Star Beer. The last program, I understand, was one of our angriest of black volunteers, a woman. She went on the air to tell what was left of the listening audience how the honkies who had set up KCHU were making a fortune off of it, and how they were at the moment leaving town, pockets stuffed full of illegal loot. Ten years later, I am still paying off the $250,000 that made it possible for her to excoriate us so piquantly over that all-too-free air.

The demise of KCHU killed some part of me as well. I had so gamely, for fifteen years, slogged about the country, dispensing beautific wisdom and media knowledge and checks so that dozens of new community stations could get on the air. I had in me this vulgar little seed called change-the-world, and KCHU, once and for all, gave me the come-uppance I needed to abandon that particular generous but destructive stance. My own brain turned dark and troubled, and I spent the following months nesting in a slightly daft mystic community in San Diego. I shuffled about, without benefit of privacy or bath, sleeping on a folding cot, learning how to make terrific bread, and wondering what the hell had gone wrong. "What is *wrong*?" I would wonder, shooing the flies away and staring at the wall. "You're being self-indulgent," some of my friends would say. That didn't explain the unexplainable—that there is in all of us a being which has great and terrible forces to bring to bear on one who builds a life on faulty foundations. One who requires the self (as I did) to function at great heights of accomplishment, without paying attention to the spirit, risks complete disfunction at some point. For months, I couldn't read beyond the first page, the first few lines of a book. I couldn't write much more than a postcard, and much less, show any interest in, or comprehension of, the world about me. At the grocery store, I would be torn between the dilemma of purchase of chicken livers or porkchops for supper. It would take a half-hour for me to make that crucial decision. Then I would have to go through the same thing in the vegetable department: corn or beans?

Self-indulgence, they called it. Pooh! This is reality. The dysfunctional mind cannot be shunted aside. Nor can it be ignored. Nor can it be brought into line by brute force, prayer, will, or hope. Inaction, doubt, self-torture, depression come to run us, and feed on themselves: thus one gets depressed about being depressed so much of the time. And then one gets depressed about being depressed about being depressed. It's very depressing, and it has an important lesson for those of us who think we're so bloody smart. That is, it's a delicate balance that runs us. There is a dark and crystalline angel within, and he (or she) *must* be cared for, recognized, answered to.

Who is to blame for the fall of KCHU? Me? Charlie Young? Dennis Gross? Larry Bolef? The City of Dallas? The local "public" radio and television media combine? The Texas Establishment? Oil Money? The Three Weird Sisters? Who knows? Who cares? KCHU left the air as slight and as slighted as it came. No one from the city, as far as I could determine, seemed too excited at its arrival; no one wondered at its departure. That has to be the real tragedy of KCHU. We, the insiders, thought that this free-forum marketplace-of-Athens voice-of-democracy was so important. We thought it could create growth and change in a city which so desperately needed criticism, self-examination, The Light from Above. Foo. KCHU was little heard and not at all loved, an orphan of the media if there ever was one. When, finally, the white noise at 90.9 was reëstablished, no one gave a good toot about what had been conceived in such hope. What elegant irony! The station died because of balkanization from within of people who cared too much. And the real important ones—the people without (all 3,000,000 of them)—saw the station as unnecessary and insignificant. That is the continuing tragedy: Dallas, Texas, huge as it is, needs a voice of freedom and doubt and wonder, doesn't it? Or does it? Why are you asking me?

Poof, it's gone. (I couldn't have poofed! so easily even a year ago: my psyche was still in bandages over this most telling failure).* The next few essays should tell you what I have come to, now that radio is less important to me. What spun away so painfully in 1977 is now seen by this bevy of scouts in my mind as but a fire denoting a new settlement, a new vista over a new horizon. Who is to blame, indeed. Who cares? For we all know (or must come to know, if we are to have any wisdom at all) that Horace was right, when he said, a scant 2,000 years ago,

Nil prepuce quiis polyester . . .

or:

Those who blame, blame only themselves . . .

December 1984

*Actually, all community stations are like the maenad. They can't be quelled. KCHU rose again (same frequency, same city, different call letters) with ACORN as the licensee.

TEACHING A PARROT HOW TO TALK
USING A RADIO LOUDSPEAKER, 1923.

A COAL MINER LISTENING TO THE RADIO, 1938.

DO CHICKENS HAVE LIPS?

I JUST got started in chickens about a year ago with a Black Cochin and a Wheaten Japanese Bantam. After that, all hell broke loose. I wrote this in a newsletter I send out to some of my friends in a Foundation we operate:

"As you know, in his dotage, your Secretary-Treasurer has taken up chicken farming. I am now the proud father of twenty-six Fayoumis, Silver Duckwing Modern, Crevecoeurs, Red Caps, Brahmans, Clean-legged Frizzles, Polish, and Mille Fleur d'Uccles. All directors will be expected to return from visits here with at least one hen under arm, and those who do not attend the required meetings will receive, Air Express, a feisty, noisy, and lice-ridden Jersey Giant Rooster (125 pounds, five-foot-eight) to boot.

"My favorite all along (I continued) has been a Black Cochin pullet, which I purchased in August. Cochins are large fluffy chickens, with feathers down the shanks and all across the outer toes so that they appear to be wearing fat bloomers. They were imported from China in the middle of the last century, and according to *The Standard of Perfection*, the Bible of U.S. Chickenry, they created a 'sensation' in the Late Victorian farmyards. As you well know, our grandfathers and grandmothers were so shy about extremities that they clothed not only themselves, but their various table, chair, and piano legs in swatches of cloth so as to protect the offended eye. Thus, Cochins were perfect for their well-dressed sensibilities.

"My Cochin sported ample dark feathers all about, especially on her rump, so that she reminded me of my own Gran'ma in black-dress-and-bustle. However, because of her large feathery feet, we dubbed her 'Sasquatch,' or 'Bigfoot.' I see her now, standing in the walkway outside, her little pea-chicken brain trying to comprehend the mystery of creatures larger than herself (me), or smaller (the sparrows who peck at her scratch-feed). At times, she will capture a fly or beetle or roach that happens by, and she will hold it aloft, athwart her relentless chaps, oblivious to the wriggling and tiggling of the creature that is about to be sacrificed as a tiny but tasty meal for my sweet Sasquatch. As she regards me, moveless, one large foot suspended in the air, her bright black eyes filled with comprehension of incomprehensibility, I see in her a mystery which belongs to all of us: that we are created by some bizarre and wondrous divinity to be as beautifully frozen in the sun as Bigfoot, the feathers reflecting back a greenish sheen in the heart of the black which quite takes one's breath away. Gurdjieff is said to have uttered: 'You haven't looked into the heart of true evil until you have gazed into the eye of a chicken,' but I do believe he would have thought differently if he knew gentle Sasquatch.

"As further confirmation of her generousness, Bigfoot gave birth, starting last week, to a series of small, brown, hard-shelled babies. At the time of incipient birth (her first), she must have traversed my office fifteen times, beak ajar, scratching in dark corners, under my chair, in the box with torn papers I so carefully set up for her. None was her satisfaction, for, moments later, the tenants of an apartment across the way (two French-Canadian construction workers who had come south for the winter) opened the door at the very moment that Sasquatch opened her door and deposited her small fry in the jamb. That's exactly what happened:

the baby was fried, small, on my grill, with butter, salt, pepper, toast and jam. It was delicious. Delicious!

"Unfortunately, Bigfoot was raised by me not to have suspicion in her little chicken heart. I lavished on her black feathers much affection and even more chicken feed so that she grew rather weighty. A big, icy-eyed German Shepherd from down the street elected (the same week as her incipient motherhood) to leap in the yard and break her neck, in a flurry of feathers, at sunset, despite my noisy objections. We laid poor Sasquatch to rest in the blue Daily Disposal Dumpster and said a small memorial service to one who perished so young, at the beginning of a great career of providing me with *oeufs fines herbes*. We assume that she has gone to the Great Silver Nest in the Sky where she is happily laying golden and tasty delights for those with more celestial diets."

I finally suggested that our Foundation (The Reginald A. Fessenden Educational Fund) establish a "Sasquatch Memorial Chair—or Nest—for all those applicants who might be intent on going into animal husbandry, preparatory to learning to diagnose those of our feathered friends who are suffering from prolapse, egg-zema, or the Chicken Pox. Or we may choose to keep some elected family knee-deep in eggs and feathers for the period of one year—which might be academic in light of my application to the U.S. Department of Agriculture for a subsidy to keep my hens in the style to which they should become accustomed, and supply them with enough mash and me with enough scratch to purchase a small army of Silver-Grey Dorkings and a hectace or two near Poway, California."

A few days after these developments, my Japanese Wheaten—constant miniature companion to the Cochin described above—came into my office (they all prefer my warm office to their cold and drafty chicken house) and snuggled in the unused nest. I tried to shoo her out, but she wouldn't move—and when I tried to push her out, I found I couldn't: she had developed paralysis of the legs.

I took her over to the County Farm Bureau where the nice lady behind the counter admired her elegant feathering, inquired as to her name ("Chicken Little") and directed me to the laboratory of Dr. Stonehenge. After I left, he put her out of her little misery, and told me later that she was suffering from Leukosis. I knew better—I knew that Chicken Little had grieved for a couple of days after the loss of her big black large-footed companion, and finally gave up and died of a broken heart.

Since then, I have experimented with raising a multitude of chicks from a variety of sources. I bought three dozen day-olds from Murray MacMurray. The cockerels and pullets that resulted weren't necessarily show-worthy, but I did get an elegant Fayaoumi which one of my friends said reminded her of a Henri Bendel model. More recently, I have taken to leafing through 'Poultry Press'—delighting in the hazy photographs and lumpy layout which reminds me less of a journal for the lovers of the *gallinaceous* and more of the "Swingers Weekly" or "L. A. Hot Dates." I cull the ads at the back and send off to such places as Ruffsdale, Pa. and Diboll, Tex. for exotic eggs. I keep hoping I will run across Mille Fleur Pyncheons and Cuckoo Creepers and Silver-Spangled Thuringers, but so far I have had to be satisfied with the White Sultans, Mottled Houdans, Barred Cochins, and White and Black Silkies. The eggs come to me wrapped in back issues of *The Diboll Repository* enclosed in Lucky Stores egg cartons. For awhile there, I was faithfully getting up at 3AM to turn the eggs in my Sears Incubator, but even a heavy dose of chicken love can turn stale at such hours, so I bought two Marsh automatic turners.

I've probably hatched over a hundred eggs, but I can't stop being excited when I see the first pip, and I watch (and hear!) the little snake-like creatures working their way out of the shell, my effort straining like theirs, until finally we lie exhausted, me on my bed, they bug-eyed on their bed of shells where they rest panting, and then careen crazily around the incubator, toppling over the other eggs and each other until I think they are going to ruin themselves on the various thermometers, wicks, bolts and supports which make up the modern equivalent of the broody hen.

It is probably an overdose of anthropomorphism, but I have decided that chicks take on the characteristics of their country origin. I had two Cornish which remind me of nothing less than a slow and dull-witted English country farmer. The Fayoumis are nervous, picky, and speak in strange clucks. The Cochins are mysterious, reserved, calm, quiet—and I half-expect their little eyes to slant up. The Plymouth Rocks show that American drive which seems too busy and over-productive.

The Sultans, on the other hand, have heavy-lidded Eastern eyes, and wear white feathery veils which obscure all but their beaks and cavernous nostrils. The d'Uccles and d'Anvers betray their Belgian heritage by being picky, competent, and busy. I suppose if I ever get my hands on some Mottled Watermaals, they would be phlegmatic and heavy and stay up all the month of October drinking and carousing on the beer I give them as a supplement to their feed. (My chicks seem to prefer Frydenlunds, a Norwegian beer, and have nothing but disdain for Millers, Schlitz, and Olympia. The Quail d'Anvers are the heavist consumers—wandering around in a constant late-afternoon fog, bumping into feed dishes and the watering cans and belching to themselves.)

I treasure the attitudes and angles of my birds. Sometimes, I catch them in frozen stances that are as poetic as a Sumi drawing. This week, I have two Belgian d'Anvers which my friend Joe is boarding with me. (He says his chickens are wild, but after two weeks on Milam's beer in Poultry Paradise Acres, they will eat out of his hand).

I can see the two Quail Belgians as I write this. Their elegant brown feathers, shafted and surrounded in gold, catch the sunlight. They are lounging naughtily in the San Diego sun like a pair of Playboy Playmates, one plump thigh rests athwart the other, feathers fluffed up, and they idly pick at the dirt with their beaks, tucking it up under their little bearded chins (do chickens have chins?)

They wriggle and throw sand high in the air with their wing bows, and if their mandibles were more mobile, I suppose they would be smiling (do chickens smile?) It's pure pleasure, enough to make me want to run out there and waddle in the sand myself, see if I can clean my own feathers a bit.

At the same time, a rather stupid White Rock Frizzle (who treats me like an invader each time I go in the hen house) turns his head up and sideways to watch me, then nips at his coverts and sickles, rolls his head back and forth on the feathers on his back. His eyes are closed, and I suppose he is as close to complete bliss as his chicken soul can comprehend.

Recently, I reread *The Egg and I*, a wonderfully profane account of chicken raising in the Pacific Northwest. It was published first in 1945, and Betty MacDonald, the writer, shows herself to be no chicken-lover: "I learned to my sorrow that baby chickens are stupid; they smell; they have to be fed, watered, and looked at, at least every three hours. Their sole idea in life is to jam themselves under the brooder and get killed; stuff their little boneheads so far into their drinking fountains they drown; drink cold water and die; get B. W. D., coccidiosis or some other disease which means sudden death. The horrid little things pick out each other's eyes and peck each other's feet until they are bloody stumps . . .

"Chickens are so dumb. Any other living thing which you fed 365 days in the year would get to know and perhaps to love you. Not the chicken. Every time I opened the chicken house door, SQUAWK, SQUAWK-SQUAAAAAAAAWK! And the dumbbells would fly up in the air and run around and bang into each other. Bob (her husband) was a little more successful—but only a little more so and only because chickens didn't bother him or he didn't yell and jump when they did . . . "

Actually, however, I did learn something about chickens from rereading this book. This is probably one of the best descriptions of the difference between a champion egg-producer and one not so:

" . . . The good layers looked motherly, their combs were full and bright red, their eyes large, beaks broad and short, and their bodies were well rounded, broadhipped and built close to the ground. They were also the diligent scratchers and eaters and their voices seemed a little lower with overtones of lullaby. The non-producers, the childless parasites, were just as typical. Their combs were small and pale, eyes small, beaks sharp and pointed, legs long, hips narrow, and they spent all of their time gossiping, starting fights, and going into screaming hysterics over nothing. The non-producers also seemed subject to many forms of female troubles—enlarged liver, wire worms, and blowouts (prolapse of the oviduct). What a bitter thing for them that, unlike their human counterparts, their only operation was one performed with an axe on the neck."

If I have to choose chicken literature, I would probably pick Chaucer. His "Nun's Priest's Prologue," written almost 600 years ago, remains as one of the finest descriptions of a beautiful rooster—Chauntecleer—for which we must draw the name of our own Chantecler* Breed. Listen as he describes the cock:

In al the land, of crowing, nas his peer.
His vois was merier than the mery orgon
On messe-dayes that in the chirche gon . . .
His comb was redder than the fyn coral,
And batailed, as it were a castel wal.
His bile was blak, and as the jeet it shoon;
Lyk asur were his legges, and his toon;
His nayles whytter than the lilie flour,
And lyk the burned gold was his colour . . .

I translate this (very roughly) as follows:

In the whole land there was no-one could crow as well as Chauntecleer. His voice was more cheerful than the merry organ heard in church during Mass. His comb was redder than the finest coral, and was ready for battle as if it were a castle wall. His beak was black—it shone like jet black marble. His shanks and his toes were as blue as lapis lazuli, his nails whiter than the lily, and his color like burnished gold . . .

I can think of no one in the ensuing centuries who has matched this lovely description of the cock.

When I was a lad in Florida, I had what we called an "English" bicycle. That means it had very thin tires—like today's

*From the Old French, *chante cler,* to sing in a clear voice . . .

racing bicycle—and all my friends made fun of me because it wasn't heavy and dull looking like a Raleigh. My Dad fixed it up with a wicker basket, and for no good reason, I took one of our five Rhode Island red chickens around with me, to the 5&10, to "Pop" Berriers, to my friend's house. She stayed in the basket without a peep.

Nowadays, I restrain myself and keep my chicks at home, in the courtyard, but still take their images about with me. The butcher, lean, with a crew cut, looks not unlike my Sultan, especially the way he turns his head to eye me when I ask for a special grind on the meat. The man on the boardwalk puffs out his chest and with his black-and-white pin-stripe suit, looks to be a dead ringer for my Barred Cochin Bantam. The sleek lady at the bus stop, all in brown, could be one of my Quail Belgians, and the wino across in the park appears like my Black Frizzle.

One of my Black-Tailed Japanese struts purposefully before me in the guise of the lady with her behind stuck out, her head pulled back, her whole attitude one of sharp disapproval. My Silver-Lace Standard Cochin—now three months old—is rangy and leggy, not unlike the teenagers that go past my front door. And the black lady in white fur I saw in the department store last week looks to me just like my white silkie.

My eyes have been afflicted by feathers, but fortunately (or necessarily) my friends are good-natured about this fascination, the newest in a long line of fascinations. After all, it can be no more bizarre than my addiction to home brewing, which stunk up the house almost as badly—or my lamentable habit of listing the ups and downs of futures, frozen pork bellies and soybean meal charts all over my office walls. It can be no worse than twenty years of scouting out transmitter sites for radio stations, or the winter I spent reading (actually reading) *Gray's Anatomy*.

Still, my friend Glen says I am sprouting feathers about my tail-section, growing red wattles; he says my nose is getting longer and hornier, and I tend to cock my head to one side as I watch him—just like Bigfoot.

He's probably right: and why not? I can't think of a nicer way to go, as a feathery Chauntecleer. The other night we were sitting about the table, several friends and I, describing the creature we would most like to be, if we had a choice. Peter Marais said he wanted to be a penguin—in fact, said that from his childhood, he *knew* he was a penguin (it's just that no one noticed).

Laura allowed as how she had always craved to be a goat, and she baa-ed a bit for us, so we could know *exactly* how she would do it. My daughter, a veterinarian's assistant by trade, sees herself as a grey, short-coated, nervous Weimaraner, and Cese, as sleek and graceful as she is, knew she was meant to be an ocelot.

Me—given my feathers and appetite for good beer—I'll probably spend the next life or so in the guise of a proud, and hopefully drunk, Bearded Silver Appenzeller Spithauben. Why not?

The Sky's No Limit, 1981
The American Bantam Association Yearbook, 1982

FISH STORY

I TOOK a day off from my onerous duties at the office to go fishing yesterday with Tom Luneau and Dave Comden. We went down to Imperial Beach—which bears the same relationship to San Diego as Long Beach to Los Angeles, Milpitas to San Francisco, Renton to Seattle, and Passaic to New York City.

We went to the Imperial Beach pier. When we rented our poles, the man at the tackle store assured us that the large-mouth bass were running that day (up to fifteen pounds he told us). He rented us three Johnny Walker casting rods with one ounce sinkers and pin-head sized fishing hooks. He also rented us four six-packs of Budweiser which we returned promptly to his back quarters as the day wore on.

It was a fine day for fishing. Those around us on the pier assured us that not only bass but trout, rainbow fish, grouper, and sailfish were abounding that day. The sun was shining, the water was fine icy blue, and the surf moderate. There was a wind from the west north west at 12 to 15 knots, and I could feel my pulse singing: "Some Enchanted Evening" in the key of C.

Dave was the first to cast out and he hooked the pier almost at once. It was hard to reel in (with his six ounce test line) but he almost had it for a moment there. The people at the rental desk were very understanding and sold him another weight and hook for less than five dollars. While this was going on, Tom managed to snag my line and there was a fierce battle to see who would end up in the bait bag.

As things quieted down, I regaled them all with stories of my Uncle Ernest who fished non-stop for forty years. A convenient job with Southland Life Insurance meant that his appearance at the office could be limited to an hour or so each morning, with the afternoons devoted to the more engrossing task of fishing for sheephead. Tom then told a silly story about his Uncle Fred whom no one liked but because he made such good Submarine sandwiches would be invited to go with the family just off Jones' Beach. Fred would reach over (when you weren't looking) and wiggle the bottom of your pole so that you thought you had a big one.

As the day wore on and the cans of beer warmed up, both inside and outside, the fishing activity became truly frenetic. Tom tried to catch my thumb on his line, and Dave snagged a truly impressive spread of kelp. Some Navy Frogmen from the Imperial Beach station sped by, and Dave assured us that they had hooked a giant White Shark. This story was naturally picked up by the fishermen all about us, and by the time we made ready to leave, the Shark was no longer than the pier and all were ready to swear they had come close to hooking the monster. The natural propensity of fishermen for truth was strained no more nor less than usual, and when Tom assured us that one day just off Long Island he had caught five mullet *at once,* and because the beer and our brains were done in, we decided it was time to adjourn, and we did so, repairing to the Squid Inn Tavern where the bartender believed all of our stories fearlessly.

April 1980

A COUPLE FOXTROTTING AT THE HOTEL VANDERBILT, 1922.

DR. SCHLESNER AND HIS MACHINE
FOR RESTORING YOUTH TO THE AGED.

DEATH IN THE AFTERNOON

LAST night a long-time friend and I went over to visit my old house on Ocean Front Walk. Ed and Sarah—the new owners—had fixed it up, filled all the planters with ice-plants, put a balcony on the back building, built a deck between the two houses, and made it a real paying apartment building. It was quite nice to be there beside the sea, on the deck, the last of the supper coals burning in the grate, the sea and the people going back and forth, the sailors rambling down Ocean Front Walk with their radios and lust intact and obvious.

My friend Jane asked me if I missed the house and the sea. "Nope," I said at once. The public aspects of the house that once delighted me now don't delight me so much—but it also may be that I don't care for the sound of the sea because in a $500,000 house, it is also the sound of the telephone ringing and various mortgagees asking when I am going to make up my late payments to them.

Ed and Sarah came out to sit with us on the porch. It was a moonless night. I felt a sudden wave of sympathy for them: for now, I am another one of the mortgage collectors: hinting, broadly, for that late payment. But ours was a social visit. It was ten or so at night. The first chill of the coming fall made me turn up my collar. The summer which had been so blastingly hot was now done for; expunged, exhausted with us as we were with it. There was a tranquility that I don't seem to remember from my tenure there.

"My dad died last Thursday," Ed said. "Just keeled over in the kitchen here. We think he was watching the sunset. I called the emergency number immediately," said Ed: "611. And I got telephone repair, and they kindly informed me that what I really wanted was 911. The paramedics were over here in a minute, blocking the alley, their sirens going. They wanted to give him CPR, and I said 'Wait. No. He's dead.' They didn't seem to understand—so I said: 'Look. He's been dead for fifteen minutes.' So they stood around, and the police came—Gary Hill (he knew my friend Holt, who used to live here)—and I said: 'He's been dead for fifteen minutes. No CPR.' "

"So we stood around and talked about Holt some, and then some kids started shooting off firecrackers on the boardwalk, and Gary went out there to shut them up—thought it was disrespectful of the dead, I guess." Ed is about forty, has one of those chins that precedes him into the room . . . prognathian chin, is that the word? He's short, looks to be pugnacious, would be a fine bantam-weight boxer. The surf was muttering in the background, and three sailors went by, and one of them gave a series of whistle bird-calls that were damn near perfect: cardinal, western plover, and something else I couldn't identify. That's one thing I miss about The Boardwalk and Mission Beach: the surprises (noise, songs, people) that go by, ten feet from one's bedroom, all night long: all night long, people with their artistry and drunkenness and problems and loves and caterwauling. "I really loved my dad," said Ed. "He was seventy years old. He loved this house, and the sunset. When he fell over, it knocked his dentures out of his mouth, so I put them back in and closed his eyes. He must've just come back from walking in the alley—we wouldn't let him smoke in the house—and I know he was looking at the sunset, and just keeled over. Everyone in the neighborhood sent their condolences—but he was seventy, and had lived a good life . . ."

". . . and went in such an easy way," I said. I was watching Ed become another person as he talked of the death of his father. I had done business with him for almost two years now. I had just known him as a man who seemed somewhat arrogant and cocksure, but a nice enough fellow at that; and all of a sudden he was telling me of the something that was crucial to life; a something that penetrated the cynical part I have in me towards all the people of all the earth. I saw someone reacting as I would want best friends and family reacting to my own demise. No 'boo-hoo' or 'Aren't you sorry for me,' but the normal tale of a human now-done-with-it.

"When the Assistant Coroner finally came, we were fairly juiced," said Ed. "When he asked me how my father had died, I said 'Maybe he was poisoned,' and the guy jumped and said 'POISONED?' until I calmed him down and told him it was a heart attack." I thought some about the Serious Nature of Death—not only the absolutely impermissible joking (as Ed and I are liable to do)—but the bureaucracy of it. Paramedics, police, the coroner. "All the time he was lying on the floor," Sarah said, "and I said 'Can't we cover him up?' He looked so cold there . . ."

"It was a good way to go," I said—but the word sounded all wrong: "Good"—can that word be used when talking about dying? I was a stranger on the porch of my own house, the porch I had built on my house beside the sea, the one that had been my home for over five years. "Do you miss it?" Jane had said, and the answer was no, of course not, for I had a few deaths of my own there, and they hadn't been as straightforward as the demise of Ed's father in the kitchen. My whole family had died in that house: Benito and Ignacio and Carlos. One day I had woken up and the whole family had been killed dead: I had been denounced as "devil" by a mother who saw me and my world there beside the sea in terms of good-and-evil, not in terms of love. My home, that place had been the home to three young men (and their dozen friends) who had raced the halls, up and down the stairs, filling my house, their house, our house with the sounds of those who are full of the energy of being young and alive and enjoying themselves, there beside the sea, where they turned tan and joyful and I could—for the years they were given to my care—be proud of the family I was feeding, and educating, and trying to teach. Trying to share my knowledge and love of books and movies and how to make it in America: how to beat the devil. And then I was the devil, and in the terrible divorce my family was no longer my family.

Sitting on that deck, next to the hot-tub, I remember Benito in this very tub, looking at the everfolding, neverending sea, Benito, of a morning. He would wake early, so early they wake when they are full of the child years, and I would come out an hour or so later to get the newspaper and he would have been there, the water up to his chin, his eyes on the level with the edge of the hottub, as he contemplated the rise and fall of the sea before him, watching the morning catching the waves come now in the sun angling in from behind him. He would be filled with the mystery of the sea in the early hours of the morning, before people and cars and airplanes and phones ringing drowned the sound of the sea: before the noise of our days intruded on the oceans. He was a boy at the edge of it all—for the three summers I knew him, this was his ritual on waking. The child of me seeing the sea. "Do you miss it?" said Jane, and I thought about missing a place that was filled with so much life. Carlos learning of Keats and MacBeth and Hart Crane. (I probably gave my "Truth is Beauty/Beauty Truth" speech in an hour-long exposition on this very porch.) Ignacio speaking jive talk for a whole summer. "Man—what'cha doin," he'd intone. "Don't you be messin' with me, boy . . ." he'd say. Benito and I just back from "The Black Stallion" and he tells me so solemnly, they can be so solemn at that age, "The movie was so good I wanted to run into the wall." All that gone in the name of The Prince of Peace. "I didn't lose Benito once," I tell my friends. "No, I lost him a thousand times a thousand times." Whenver I go to a movie alone (how he loved Charlie Chaplin); each time I cook a great meal alone (he loved my cooking); each time I play Pac-Man. The loss gets factored (as they say in Math), and it comes to me over and over, sometimes in waves, sometimes in isolated squalls. What is that Hasidic saying?—if we knew the grief of people, each human's real tragedy—we would never be angry with anyone, never shout, never be insulting, never hurt anyone willfully.

The love I felt for that family has been replaced with a . . . a what? "It's an ache . . ." I'll say; "It's a feeling of emptiness . . ." What do those phrases mean? "Words don't convey emotions," one of my dreams told me: "They just tie them together." We grow up to believe in words as conveyors of our feelings: and yet, when the time comes—they might as well be stones. The Atacama Desert filled with the stones of our grieving, all of us grieving over all our losses. "Give us this day our daily stone . . ." said Miss Lonelyhearts.

The Benito and Carlos and Ignacio I knew and loved from back then have long since disappeared. They've had three years outside my purview to grow and change. The boys I raised are dead. The sight of great gold waves coming in, soaring like banners, inspired in Benito's eleven-year-old mind the same wordless awe with which I now view his absence. I will now, as I did then, honor his silence. He no longer exists. The sun dies over the dark sea; and, in the sea's voice, I can hear the bumping together of dark bodies, bodies out of my memory.

"We love them too much," I think. It must be against the rules to love them too much, for they come at night and steal them away. And then you want to run into walls. Or into the sea. You want to destroy people, people who use the sweet words of Jesus to tear others apart, tear apart others who think differently about the gods in our lives. "I had an operation on my brain," I say to friends. "Since I couldn't stop thinking about Benito, I had this surgery done to take out all the circular, repetitive, parrot coils in my brain. You should come over and see my hippocampus. I had them put it in a jar, and it sits on the mantle. Sometimes I catch myself looking at it and I think: 'Hm. That brain, on the shelf. I wonder whose it is, what it's there for . . .' "

"Everyone must lose someone they care for too much," I tell myelf. "It's the human condition. Life isn't permanent. Look at all the people who have died, leaving someone bereft," I tell myself. "Just be glad he's not dead," I tell myself. His birthday will fall in exactly a month. Already I have composed the letter he will get from me (she can't deny me a message on his birthday, can she?) I will work on it more, I know: changing a word here, deleting there, making it, how do they say it, letter perfect. He can't read he can't read but I still have to send him something, some momento of our time together. I don't want him to begin to think for a moment that I have forgotten him.

My Dear Benito:

Last night I had a dream about us. I dreamed that we were walking through the dark forest. The trees were black and filled the sky. Ahead of us there was a clearing, where the moon was starting to rise. You and I seemed to float over the surface of the land. There were a few red-eyed black-winged birds, flapping noisily above us.

When the moon came up, its face was filled with the image of a double-headed eagle. What had happened is that someone on earth had invented a laser that could project all the way to the moon, project pictures on it. When we looked up, the first thing we saw was the picture of the black-and-white two-headed bird. Later when the moon had risen fully, and filled most of the horizon, they began to show movies . . .using the face of the moon as a giant screen!

They showed old time movies—Laurel & Hardy, Buster Keaton, Charlie Chaplin. How fine it was that they had decided to use this huge round screen in the sky for silent comedies. The whole of humanity every night would be able to see movies that would make them laugh. No politics, no hate, no violence: only the noiseless comedies of the past, projected for the 10,000,000,000 eyes of the world. A peasant in Egypt, a Sikh in India, a dirt-farmer in Ecuador, a trader in the Congo. Each night, they could see high comedy—for free. You and I were laughing and laughing at the cleverness of it.

I think of you every day. I remember the times when you used to wake early in the morning, and steal out to the hot-tub, and sit there, silently, watching the great sea rise and fall, rise and fall. You would stay there in the hot tub, your hands crossed before you, thinking whatever Benito thoughts you would think, silent before the great majesty of the sea. I would get up at eight or nine or so, and you would still be there, a part of the sea, and yet alone. I remember thinking at that time that I wished I could be inside your head, with your thoughts. Inside your head.

I look forward to that time when we will meet again. I say to myself: 'Let's see—this is 1984; he turns fourteen this summer. That means—five six seven eight—I'll be able to see him again in 1988. Or is it 1990?' My mathematics was always terrible. 'I hope he doesn't forget me,' I think. It's so hard to try to tell people to remember you, after you've been apart for a while. I always catch myself trying to imagine what you look like. 'I know he's tall,' I think. 'I wonder how tall he is now.'

One day about three weeks ago I got very sick. 'I'm going to die,' I said to myself. I had dreams about computers, about getting caught in the machinery of them. My fever went to 104. My neighbor Tom came over and worried about me, talked about taking me to the hospital.

"To hell with that, Tom," I said. "If you took me to the hospital, I would just die there, and have none of these pictures to think about, to engrave on my memory." I was talking about the fact that in my room there are pictures of Benito from back before, before . . . Benito climbing trees, Benito walking the beach, Benito turning his head to look at the camera, in black and white. "What's going to happen," I told Tom, "is that they are going to invent a laser that will project pictures all the way to the moon. That's why I have to wait (I can't die now) until they have it ready, and then I will project pictures of my great and good friend Benito onto the moon, so that everyone can come to know him and to love him as I have. I can't die now," I said. "You can't take me to the hospital," I said. "We have to throw his picture thousands of miles into the sky, so everyone can get to know him . . ." I mumbled on to myself, passing out in the midst of my feverish delusion. "We've got to let the world know about him," I kept mumbling, as I fell asleep.

And so I fell to my troubled sleep; but I remember thinking: "Happy 14th birthday, old friend . . ." And when I woke up I was well . . .

With Affection,
Lorenzo

"He loved the sunset," said Ed. "No," I think, "he loved the sunrise." So amazed at its force that he was stilled, at the first few hours of his young and lovely life. The boy in me, the one of the knobbly knees and the wry smile, knocked into unaccustomed silence by the morning of our days. We love the sunrise; its peace is a special vortex about us: and then comes a storm center (you never know when the winds will come) and you treasure the peace of those waves arching in so languidly over the gold-come-morning, ignoring the storms lurking further down the beach, coming inland at such a desperate speed.

"You'll have to come back and cook," says Ed as we are leaving. "After all, you know the kitchen as well as we do. You know the house as well as we do." "Of course, of course," I say: "Come back and cook." I do know the house—maybe even better than Ed does. "That's a good idea." I say, but I'm not going to cook; I won't go back. I told him I was sorry about his dad, and Jane and I left, but I don't think I'll return to the house of the dead. I prefer to leave the dead to the dead. Any attempt to resuscitate them we leave to others. Leaving the raising and settings of suns to the others.

September 1984

CLARA HORTON LISTENING TO THE RADIO, 1925.

LEARNING TO DRAW BY RADIO BROADCAST, 1924.

TIJUANA BABY JESUS

I LIKE it best in the evening, when the Gabacho sits on the porch, and gets drunk, and he starts singing, or reading out loud. He'll read poetry to us, in his drunken, deep voice, with its gringo accent. I always want him to read the one about the wall:

I shut the door of my balcony
Because I don't want to hear the weeping
Here behind the thick grey walls
I can hear nothing but the weeping.

The word for "weeping" is "llanto." It's one of my favorite words:

There are very few angels that can sing
There are very few dogs that can bark
A thousand violins
Fit neatly in the palm of my hand.

He always talks about the Tijuana dogs. He says they are just like all the animals in Tijuana—and all the people. He says they have to be tough to survive. He can never understand why we are always throwing rocks at the "pinche perros . . . "

The weeping is an immense dog
The weeping is an immense angel
The weeping is an immense violin.
The tears muzzle the wind
And you can hear nothing but the weeping . . .

He doesn't exactly read it: he sings it. And sometimes I wonder if he knows what the words mean. He says it was written by a Spanish poet, who was killed during the Spanish Civil War when he was still very young. "Every country has to have a civil war," he says: "like the United States in 1862 and Mexico in 1914 and Spain in 1936. That's the way they can kill off the boys," he says, "like you and Martin and Mundo. They'll come to get you too," he says. "They kill off the boys. The old men can't stand the boys." He waves his beer bottle at me. "The politicians can't stand you. That's why they'll send you off to war. 'There are just too many boys,' they think, so they start another war just to put you in the ground, make you an unknown soldier." He talks like that a lot.

We live in the Los Cipres district of Tijuana. It's 'way outside the center of town, towards the beach. It's right next to the border—three hundred yards south of the fence that they built to separate the two countries. Sometimes I go up to the hill that overlooks the border, and watch the sun set. I look at the dividing line that cuts between the two countries like a knife. I watch the long shadows that stretch out from the fence, stretching out like fingers. The shadows around the fence turn the color of ashes. I don't even know why I am sad when I look at that fence, cutting between countries—but I am. Maybe it's because it tells me that we'll always be separated. Or maybe it tells me that just over the border, there are boys like me, boys who will never be able to cross over, so that we can talk, so that I can get to know them. It's not just that we speak different languages. It's that they live over there, and I live over here, and we can never reach each other. I see them running, or playing baseball—I watch them through the fence—and I know I can never run with them. There are people like us on both sides of the fence, fences all over the world, separating people, and all we can hear is the sound of the weeping.

I think you would like him if you met him, the Gabacho. He's tall, and has these crutches he walks with, and has eyes that hide behind his thick glasses. Until last year, he lived on the other side of the border. He lived in a big house with eight rooms, right beside the sea. Each morning he would wake up and look out the window and see the ocean turning from black to dark blue to blue-green to green as the sun came up. He would lie in his bed in this big house with the polished wooden floors, and he would listen to the sea talking to him. The waves crashed over each other, and onto the beach right outside his window, and he listened to what they said to him, and one day he got up and closed all the windows, locked the door, and put a sign on the house that said FOR SALE. He moved down across the border to the two-room shack here in Los Cipres. He could have gone to one of the gringo colonies—with all the tired white-faced Americans—but he came to Los Cipres instead, to the house where Doña Martina used to live, before she died of cancer of the stomach. He moved into her place, with the cement floors and the outhouse and the flies on the ceiling and the ants. When he moved in, he brought his suitcase, and a big bed, and a few books, and a gas stove, and five office chairs to sit in.

Once I asked him why he moved here, where the flies wake you up at six in the morning by trying to crawl in your ears, and the water trucks can't get through in the winter because of the slides and the storms—and he said that he moved here because he was suffering from a broken heart, and the only people in the world that understand a broken heart are the Mexican people, because people have been breaking hearts here for seven hundred years. He said he knew if he moved here everyone would leave him alone until he could stop thinking about his broken heart. That was a year ago.

Sometimes, the Gabacho will take us shopping. We go to the public market in Rosarito Beach, and the Calimax. He'll make up a list for us.

jabón	soap
pan	bread
cebollas	onions
ajo	garlic
gusanos	worms
vinagre	vinegar
mantequilla	butter
sapos	toads
mopeador	mop
girafes	giraffes

He'll give the list to me, because he knows I can read, and I'll be going down the list, looking for the garlic and onions, and then I'll get to 'worms,' and I'll look around, trying to figure out where they keep the worms, and then I'll see him looking at me, laughing, because he knew just what I was going to do. It's a joke he plays on us.

He walks with his two muletas—crutches, only he calls them "maletas"—which means "suitcases." "Dáme mis maletas," he'll tell me. "Dáme MULetas," I'll say to him. "MALetas," he'll say. "MULetas," I'll say. We'll go on like this for a while. He has steel rods on his legs, going down to his shoe, and a "faja" for his waist. He had polio when he was eighteen—more than thirty years ago, before I was born. He said that he had to stay in bed for two years, in a charity hospital, and then had to learn how to walk all over again. He says he stopped being a boy then. He says that's why he likes being with us, because we are where he was then. He is almost fifty years old now. My mother is fifty years old too—but I think Gabachos and Mexicans grow old differently. My mother's hair isn't dark like his, and her skin is lined and wrinkled from worrying about me and my brothers and sisters and cousins—the twelve of us. She sometimes wonders what we're going to eat. He doesn't worry about things like that.

He'll ask me, the Gabacho will, to name off all my brothers and sisters. "Well," I'll say: "There's Paco, we call him Bibi; and there's Francisco—that's Kiko. Then there's Angela and Tina and Miguel and the little one Consuela. There's the cousins living with us now because their father, my mother's brother, got shot one night crossing the border near San Ysidro. So we have Lina and Marta and Lupe and the little idiot Jesús." I stop for a minute because I can't remember them all. The Gabacho will make me add them up, but I am always leaving out someone, and getting the numbers wrong. There's always someone I'm forgetting—sometimes it is the little Jesús, and sometimes I forget Salvadore because on weekends, he comes in drunk and hits me. And sometimes I even forget me! You'd think it would be easy for me to remember, especially since we are all sleeping together in the same room, the one that used to be the paved-over courtyard.

Sometimes, when the Gabacho is home all day, and I'm there all day, eating breakfast, and lunch, and supper, and fruit, and cereal, and drinking a beer or two—he'll ask me if it's all right that I stay with him there all day and most of the evening. "What does your father say?" he asks me. My father? What does he know? He's up in Salinas, picking vegetables, if he can get work. My mother—she thinks it's one less mouth to feed tortillas and beans. She says I'm too lazy anyway, sitting and looking up at the hills and dreaming about what's going to happen to me when I fly north. She says that with him paying me 1,500 pesos a week, I'm better off at the Americano's house—"Maybe he'll teach you how to work," she says. She doesn't know him. He just pretends to hire us to work for him. He really wants us to keep him company.

He likes to cook, the Gabacho does, and he'll cook norteamericano dishes. Pork chops with rice and cumino, or roast lamb with lemon juice, and rosemary, and slivers of garlic stuck under the fat. Beefsteak rubbed with olive oil and sprinkled with fresh pepper and garlic salt. He makes me cut up all the things we put on the food. That's how I know how he does it.

Suppertime is the best time at his house—there's always five or six of us, and he is drinking beer, and laughing, and playing the guitar, or playing chess with Wini, or asking us questions: "What does your father do?" "How much money does he make?" "How does he get across to the other side?" "When does he come home?" "Where did your abuelita come from?" "What do you eat for breakfast?"

The rooms in his house smell so rich with cooking garlic and onion. Sometimes I wonder what the ghost of Doña Martina, the woman who lived here before she died, thinks about us there in her house, with our jokes and singing and telling stories and eating all that food. The house was always cold and quiet when she lived here—there's a sign on the door, which the Gabacho left up, that says "Esta casa es hogar católico"—this home is catholic. She was very serious, and tried to teach the children in the village the catechism, and rapped the children on the knuckles with her cane when they didn't learn their lessons. She was so serious, and now the house is so full of people singing and teaching the Gabacho dirty words and fooling around. The Gabacho always likes to light four or five candles, stuck in wine bottles, for us to eat supper by. Sometimes, I wonder if it's her ghost that blows out the candles, blows cold air on the backs of our necks. Whenever that happens, I move closer to the Gabacho. He says that he doesn't mind spirits—in fact, he says they are his friends, and sometimes he talks to them when he's lonely or trying to get to sleep. I think he might be saying that just to put a fright in us. And it works—me and Martin always move closer to him when he says that.

When he puts the meat on the table, he'll always have some vegetable, like broccoli, or cauliflower, or artichokes. He makes us eat vegetables because he wants us to grow up, not out. So he'll make me peel the broccoli, and heat up butter in the frying pan, and squeeze two or three limas in it. If I leave it on the heat too long, it'll start to smoke, and the Gabacho will call me a "pinche cabron", and a "knucklehead". I know what a "pinche cabron" is—but I don't know what a "knucklehead" is, and if I ask him, he'll say he can't translate it. I'll ask him again, and he'll say "no entiendo," I don't understand. He says that all the time. I'll ask him to take me to the swap meet, or I'll ask him to buy me another bicycle to replace the one that got stolen by Wini's brother, or I'll ask him for a peseta for a Coca-Cola—and he'll say "no entiendo," no matter how many times I ask him. He understands what I want, only he likes to hide behind his language.

The Gabacho says we all speak bastard Spanish. He lived in Spain, in Andalucía, for two years. He says they speak a better language over there than we speak in Tijuana. He calls it "our bastard Spanish—Pinche Tijuana Español." He says they don't call a car a "carro," or a truck a "troque." He says that when they hitchhike—they don't call it "pita un ride . . ." or call junk "yonque." He says our Spanish is a disgrace and a shame to the Iberian peninsula, and that it should be left on the doorstep of the orphanage. He says the conquistadores would be offended by us calling a veranda "porche" and underpants "shors" and a sink a "sinke."

He has a few favorite words. Like "vosotros" which we never used until he came here—it means "you." He says we should get in the habit of using it, but Martin said that if he called his friends "vosotros," they'd never know what in the pinche hell he was talking about. The Americano likes the word for water bottle, too—"garafon"—and especially the one for liver—"hígado." He says that's the funniest word in the Spanish language.

His favorite sounds in the village are the water trucks that go by in the morning, with their mournful horns, and their powerful roaring. You stand at the window and hold up one finger or two, for one or two garafones of water. He thinks it's more friendly than having your water sent through a tube by a man you've never met, who you can't talk politics or economics with.

He's crazy about chickens. He brought some funny-looking chickens to the village—ones with feathers down their legs, or ones with white capetons on their heads. He says they are 'pollos polski'—Polish chickens. His favorite is a rooster he calls Walechsa. He says his favorite thing in the morning is listening to them sing "Ki-Kiri-Kee!" Of course, they don't just crow in the morning. They do it all night long. He says they think they have to make the sun come up—and, being Polish chickens—they crow all night just to make sure that the sun will come up.

I think he likes chickens better than dogs, or cats, or even babies. He says one of the main national products of Mexico is babies. "Dust, tortillas, and babies—that's what this pinche country is best at making," he says. He thinks that if people in norteamerica had any sense they'd have families with twelve or fifteen children just like we do. He says it keeps you busy until number six or seven, then the older ones can take care of the younger ones. He calls it a "perpetual people machine."

I never know when the Gabacho is joking. Like when he's talking about the Tijuana flies. He says you can tell the American flies from the Mexican flies. He says the Mexican flies always eat things that the norteamericano flies would never touch. He says the Mexican flies always come around at night, and never have any identification. He says that if you stand at the border, you can see them coming back, in the dark, with all that food strapped to their little bodies. He says it's just the opposite of the Mexican pigeons: all the smart ones go north. They know that if they stay here, they'll get eaten—so they either have to wetback it over the Tijuana river, or change their names to doves. I told him it didn't make any difference, since the word for pigeons and doves is just the same. He says "Sí—como he dicho." That means: "just like I said all along," which is not it at all.

Sometimes, the Gabacho will take us to Rosarito beach, to the swap meet. He pulls himself out of bed, and puts his pants on, and we get his leg for him, and he'll strap it on, and he puts on his corset, strapping it up, and then he gets up on his silver crutches. He walks very slowly on his crutches, and you have to be very patient to go with him. Negro one time told him that his crutches weren't really crutches, but wings. Negro says that someday the Gabacho will be able to fly with these silver wings. Negro says he'll turn into a bird, and these will be his new wings, and he'll be able to fly with them, just like a great hawk, one of those you see hanging motionless over the valley: the ones they call "kites." It's a bird that hangs up there in the air, and then it sweeps down and catches a rabbit, and pulls it up in the air, wriggling, into the air. Negro says the gringo is like that—with his silver wings and his nose like a beak. "And we're the rabbits," Martin says. The gringo thought it was very funny, Negro calling his nose a "beak."

We drive to Rosarito Beach in his van. Before the Gabacho came over the wall, I had never been in a car—just the "burros," the busses that go around Tijuana, and they never go fast. When he's driving, he likes going as fast as he can, and you hang out the window, and open your mouth, and the wind makes your cheeks puff out and your eyes sting, and you can move your hand up and down in the wind like a snake, and you never want to stop going.

After the swap meet, he will take us to the Cafe Poblano, where they have the six tables and the old man with baggy pants and telescopio eye-glasses. The old man comes over and puts his hand on your shoulder, and asks you what you want to eat, and the Gabacho will have his first beer of the day. We'll eat tripe soup, or pozole, or carnitas, and the Gabacho will be drinking his Corona and looking out the window. He won"t even be listening to us, paying any attention to us. He'll be looking out at the ocean, thinking of something, maybe the other house he had to the north, the one he gave up because of his broken heart. He'll watch the waves, and think what it was like there to the north, to wake up in the morning, to know that the great sea is out there, stretching 10,000 miles to the west, stretching on half-way across the globe—with the thousand thousand waves curling and breaking across the face of it, knowing that it's just like a giant heart—the great open heart of the world, with these tremors running across the surface of it. There can be a monster earthquake going on somewhere down below, but you would never know, because it is so big, the great sea never shows what lies underneath it. It just keeps on rising and falling, never showing, not for a moment, the earthquakes in the depths of it.

I think it's Sunday mornings I like best. The Gabacho has a huge bed, on stilts—it's the biggest bed I've ever seen. I know that me and Conchita and Francisco and Jesus and my mother and all of us could sleep in it and never wake each other up. The Gabacho stays there all morning, drinking tea with milk in it. It looks like slop you'd feed to the pigs, but he drinks three big measuring cups full. He writes in his journal—he'll fill ten or fifteen pages with his scrawl. I ask him what he's writing, and he'll pick it up and look down his glasses and read to me: "My Uncle Carl used to eat little boys. He'd cook them up in pot-pies, with oregano, parsley, pepper, salt, and cream sauce. He would bake the pies at 350 degrees for 45 minutes, until they were crunchy brown, and then he would take them out of the oven, and sit down at the table, and cut into the crust, and start eating on their legs like this . . . " and just when he says that, he grabs my leg and starts to bite in on it, and I have to pull it away from him, because it looks like he's going to eat me. I don't think that's what he's writing in his journal.

It's Sunday mornings I like best. I'll lie there on the bed, next to him, with Martin and Elfren. He'll be writing, and drinking his tea, and looking out the window—and sometimes he looks out the window so hard and so long, that I think he's reaching his words from out there. I'll look over my shoulder, to see what he's seeing, but all I can see is the cottonwood and the sage, and the hill across the valley where the rocks have spilled down the arroyo from last winter's storm, a red and yellow wave across the face of the hill. We can lie there for hours, and at times I curl up next to him, and fall asleep. He listens to that strange, boring music on the radio—he calls it Bok—Yo Jan Bok. We'll be nodding off to sleep, as if we had always been living there, with the music, and the sun coming through the window, lighting up the room, making it warm, and our troubles are so far away, our families crowded together so far away, and I think I want to live like the Gabacho, have a big bed, where I can lie on Sunday mornings, and write in a big book, and listen to this Bok.

Sometimes I try to think of what the Gabacho was like before we knew him. I can't imagine he even existed before he came to Los Cipres. I think somehow he has always known us, always been here. Maybe he knew us from some other place, a long time ago, when we were all different people. I can't even believe he *was* before he met us. I think that he just started existing the day he drove down the hill for the first time, into the valley. It's hard to think of people before you knew them. I asked him what he was like before he came here, and he said that before he came here, he was scared of dying. He wrote down for me on the blank page of his journal

TIMOR MORTIS CONTURBAT ME

which he said meant "I'm scared to death of dying." He said that when he was living to the north, he was starting to die, and he didn't like that, because it scared him. He's the first man I've ever known who says that he is scared. None of my friends ever say that. If you have to fight with the biggest guy in school, you stand up and put up your fists and try not to shake, because you don't want anyone in the school-yard to know that you are scared.

He said that for him being in Tijuana was like looking in the mirror. He had always heard that the Americanos who lived outside the beach communities would get beaten and robbed. He said that he had heard that you could get put in jail, and never get out again. He said that he had to come here because it scared him. He said that the things that scare him the most are here in Tijuana. He said he wanted to get to know his fear better. He said it was like an old friend. He says the thing you fear the most is the thing that will get you in the end. He told me that if you're most afraid of getting beaten up—then that's what you had to face: go into it; become the thing you fear. He said he had to do it. That would be the only way our lives would be worth anything. He said most people were scared of dying; so naturally that was the thing that would come to all of us.

What he was saying reminded me of El Milagro—the Miracle. I go to the cathedral here in Tijuana with my mother, and it's dark, and you go in, and look up at the Holy Mother, and she's looking right at you. You can see she's breathing, because her chest is moving, up and down, up and down. Her eyes are wide and dark, and she's looking right into you, into the deepest part of you. You can see her lips move, and she's calling out your name. And when you look down at the holy babe, you can see he's staring right into you too. His eyes are wide and dark, darker than hers, baby-dark eyes, and he's staring into your heart. His baby hands are moving, and you know they are signaling to you in some strange way, and he's asking you with his hands that you show him your heart, but you are afraid, because you are afraid he will see all the dirt and ugliness in your heart. His eyes are as deep as any well that you've ever looked into. The church is dark, and here are these statues moving and calling out to you, and you want to tell your mother, but she would never understand. I told the Gabacho about that, that thing that happens to me, what I think of as the Miracle of the Tijuana Baby Jesús (that's what I call it) and I know he understands. He said that something like that happened to him when he was a boy. He would look at a picture of his grandfather. His grandfather had died the same day he was born—and he had the same name, too: Lorenzo. He said that if he looked at this picture long enough, the lips began to move, and he knew that the old man was calling out to him, calling "Lorenzo," "Lorenzo." He'll tell you stories like that, and won't laugh at you if you tell him your stories, like the one I told him about Tijuana Baby Jesús.

One day the Gabacho left and never came back. I don't know what happened to him. No one does, because he left, just like that: left all his furniture and plates and books and chairs, and went over the border and never returned. Some people say he got scared, or tired, or bored . . . or didn't like the people in the village. Half the people in the village owed him money. They always borrowed money from him—but maybe that is why he left—because people might not have wanted him around, so they would have to pay him back. Jose Carlos said the federales were after him, but I don't believe that. I think that one day he woke up and his broken heart was cured, and that's why he left and never came back.

Anyway, he's gone now. We miss him: the rides in his car, and the Sunday mornings on his bed, and the suppers, with candlelight and broccoli and roast lamb and the funny stories and the songs. He was my friend, as much as any one can be, and I think of him, wonder when he'll come back, driving back in his great orange van, with all the bags of groceries and presents for us and the newspapers and the peaches, and pears, and grapes.

Sometimes I go to the top of the hill, overlooking the border, and look for him. I stay there all day sometimes, hoping to see him somewhere on the other side, coming here, coming down here, to come back to us again. Martin thinks he's dead, and José Carlos says he's in jail; Negro thinks he found a woman and fell in love and got married, and Mundo says he just got bored. I don't know—I don't think he would leave us without a good reason, without telling us.

I look to the north, to the great silver buildings up there, on the horizon, shimmering in the sun—buildings that I've never seen up close, buildings sprouting like trees, with silver bark on them. I see them, and then I see the fence, cutting across the land like some great wound. "He's somewhere up there," I think. "Somewhere, he's cooking, or writing in his journal, or drinking beer: maybe looking down here, to the south, where he used to be so happy." I watch the green trucks of La Migra going back and forth: the migra, with their red faces and their guns and their hard eyes. They go back and forth all day, all night, back and forth, keeping me from going north to look for him, my friend, the Gabacho. And sometimes they fly over with their helicopters—with their red flashing lights, and going "whomp whomp whomp"—they swoop down with their searchlights beaming in on the Mexicans who are trying to get across what they call the War Zone. At night you can see the searchlights playing across the trenches, and you can see the shadows running across the barren land. Sometimes you will see the helicopters swoop down, and the "pollos" will jump out of the holes where they are hiding, and begin to run, and two or three of the trucks will race over, and you'll see La Migra jump out with their guns, putting our people in their trucks, taking them away.

I wish the Gabacho would come back. I'd ask him who made the border, who made the fence that runs across our valley. I'd ask him who says who has to be on one side and who on the other. I would ask him what it's like to go north, to be able to travel freely, without fear of being caught and sent away. I would ask him why they have to run after people like that, with their trucks and their guns and their helicopters—I would ask him why they have to do that. I think the border is sad. It's not a person, and it's not a country, really: but it's sad just the same. It's like cutting yourself with a knife; you do it, and you look down, and all of a sudden you are sorry, because you didn't mean to hurt yourself. The border's sad like that.

When I was walking
Along the belt of the earth . . .

he used to read:

I saw two marble eagles
And a naked girl.
One was the other
And the girl was nothing . . .
Through the branches of the laurel tree
I saw two naked doves
One was the other
And the two were nothing.

That part about the naked girls makes me think of the Mexican girls. I want to go over the border for them. I'll go over there and find the Gabacho in his eight room house, and he'll give me a job, and I'll work hard for him, and save my money, and become rich like him. I'll be able to live like the Gabacho, and save up lots of money. I'll come back some day, rich, my pockets full of money, and I'll find a girl, a Mexican girl, one I can have for a 'novia.' The Mexican girls, their faces and bodies are so full of love. They aren't cold and polished and piled up like American women, all thin and icy. No—the Mexican women are alive and moving and soft and easy. They walk so beautifully, tanta buena—so that you *have* to love them. They are proud, and beautiful—and after I get through working for the Gabacho, I'm going to come home with my pockets full of money so that I can love the Mexican women and they can love me.

I've thought about that a lot since the Gabacho left. I think about finding him, and working for him so I can get the money, so my mother will never have to work again, scrubbing with a scrub board. No—I'll buy her a washing machine, and my brothers and sisters will never be hungry again, and they'll have clothes that fit, clothes that aren't faded and worn. The Gabacho will help me to go across the war zone, and I'll work and be able to buy a car and come back to the village, with a car, and nice clothes, and I'll look so good that the girls will want to go to the disco with me. I'll get a dozen pants, and leather jackets, three of them, and six pairs of boots. I'll put the whole family in the car, and we'll drive down to Puebla, where my grandmother lives. She'd be so surprised to see us. She'll ask us where we got the money, and they'll point at me, and I'll be so proud, proud that I could work for the Gabacho and do something for my mother after all she's suffered for us.

There's something else I'll do with the money. I'll get a present for the Gabacho. You know what I'll get for him? I'll go all over the world, and I'll find the best doctor in the world—whether it's in Los Angeles or New York or Miami or London or Mexico City—and I'll buy the injections he needs so that he can walk again, without his 'silverwings,' I'll find a doctor who will cure him—surprise him with this doctor who will give him the injections—and then, afterwards, he'll be able to hang his silver wings on the mantle. He'll look at the "wings" any time he wants, and know that he flew with them for twenty-five years, but he'll know that now he's free, and doesn't have to use them any more.

I want to do that for him because of what he's done for me—and because of the dream. The dream: it was such a strange dream, but it was so clear—I can see it right now, as I am telling it to you. It's a dream about me and the Gabacho.

He's walking—just like me and Negro and Martin. He isn't a cripple any longer. He's walking with these giant strides, and he's big. He's as big as one of the Laurel trees down near the road. He's so big he can walk over the mountains, and highways, and rivers.

In this dream he has me on his shoulders. We are going across the land. The mist is rising up—sometimes it comes all the way up to his shoulders, and I can't even see my legs, like they were invisible.

We are walking across the land, and it's so rich and fertile and full, we can smell it, just like the vanilla beans that my grandmother used to buy for me in the public market down in Izucar de Matamoros . . . the vanilla beans that smell so rich and strong, and she puts them in the hot milk, when it's Los Reyes— and it smells so good you don't want to drink it. That's how the land smells.

I am holding onto his chino hair as we are walking across the land. We are talking: I am laughing and whispering in his ear. I can tell him which way to walk, if I want to, and he'll do it. Maybe we'll be singing, the song he used to sing late at night, when he was drunk, in the evening, on the porch—the one about the "lovely senorita . . . "

La buena senorita dice "no"
Pero no quier salir
Hasta me diga
Quizás, quizás, quizás . . .

We'll be singing that as we cross the valley of Los Cipres, and I can look down from way up there and see Martin and José Carlos and Negro and Elfren running after us. We see them trying to keep up with us and pointing, they look so tiny down there—and they try to catch us, but they can't, because we're going too fast, and the wind is in my hair, the wind rushing through my hair.

When we cross over the border, the land will suddenly become shrunken, and hurt. All the trees will come to be tiny, and bent over, and twisted. Blood will leak from the branches, like blood from the finger-tips of a dead man. There will be hundreds of lizards, and geckos, and gila monsters, and there will be a bird that flies right up in my face, trying to get me. It is a huge bird, with a body the color of lead. Its face isn't the face of a bird at all, but the face of a baby, one that has died a horrible death. Its mouth is twisted open, its tongue is sticking straight out, and the tongue is black. I am scared that it's going to catch us, and I beg the

Gabacho to hurry, so it won't get us, and he outruns it, and we are striding over the desert, up to the top of the mountain. The sky is white, like it was made of cotton, and there is nothing in it except for the sun coming out of the sea: alone there, the sun, at the edge of the horizon, the sun a great circle, with big lids, like an eye. It is rising up out of the hot black sea, growing into the white cotton sky, and when it finally escapes from the black water, and gets into the sky, it stops moving. As soon as it sees the Gabacho with me on his shoulders, it stops, and looks at us, and we look at it; and then . . . it winks at us. A great big cool wink—just like it knew us, just like we were old friends, like it has known us all along, and will know us forever. That's why it winks, because it knows that we're part of this big secret joke, a joke between the Gabacho, and me, and the sun. We're old friends meeting there, on top of the hill, with our joke that no one else in the whole world understands, except us: me, and the Gabacho, standing so tall there, with the sun come up so fresh now, the sun now come up . . .

January 1985

BUSTER, 1929.

THE END

I'm in Seattle, staying at the Sherwood Inn. Room 202 is right next to the boiler room. All day and all night there is a throbbing, as in a great ship. You remember how it is: you're on one of those liners en route to Europe. The whole boat rumbles and throbs with the giant motors.

Each night at the S. S. Sherwood I go to sleep with the sound of heavy deep rumblings in my head, rocking my bed. Steaming across the valleys and mountains of Chehalis and Yakima, taking me to Le Havre, sailing to Byzantium.

I am up here for the closing of the sale of KRAB. It takes place on the 13th of April, in the conference room of Wheedle & Dum, old-line Seattle law firm on Third Avenue. The buyer is Sunbelt Broadcasting, and its owner is balding and nervous looking, surrounded by attorneys and accountants and attorneys. His communications attorney comes in and smiles at Mr. Sunbelt and snarls at us for awhile. The usual closing sabre-rattling . . . every sentence designed to imply *if you don't like what we're asking for, why we'll just kill the whole deal*. Heady threats for a process that's been in the making for over a year.

Security Pacific Bank is there: they're financing the sale. Their bank officer looks to be a parody of every California hot-shot finance officer—salt-and-pepper hair, salt-and-pepper suit, salt-and-pepper smile. He keeps washing his hands in Jergen's Lotion kept in his $499 briefcase. Greasing his palms.

Mr. Sunbelt spends some four hours absently listening to the wrangling fly-buzzing of the lawyers, and signing papers. Notes, contracts, agreements, sub-agreements, supra-agreements, amendments. His attorneys mumble over papers, nod and shake their heads, proving that they are earning their $150/hour fees. He must've signed his John Hancock two hundred times that day, and I wonder if he was ever tempted to sign

John Hancock

to see if anyone would read those papers, ever. As many pages as went into *War and Peace*, filed away in various attorney and accountant and banking and government offices, testimony to a day in mid-April in Seattle, closing a twenty-two year old station, closing a period of my life.

I catch myself wondering what those people thought of the rather seedy collection of misfits at the Southern (or weaker) end of the Board Room. Margason, obviously a radical *manque* in his jacket no older than the station. Milam with his ratty shoes and frayed Brooks Brothers shirt. Bader—who disguises his all-too-prosperous firm by affecting dun-colored suits from the 60's.

It's hard to be restless in that room—although the meanderings of threats and scratching of pens go on for hours. Economic determinism: the room reeks with it. What was once a small and silly station in an obscure corner of the United States has come now to represent something far different. What was a statement (political, social, economic) in 1962 is now a vast resource called Millions-Of-Dollars. What was Margason's and Milam's and Lansman's fond hope for change-in-communication is now something completely different: certainly as controversial, certainly as painful.

KRAB grew out of rage at government temporizing a quarter century ago. It was two years in the gestation, and twenty-two on-the-air. Our history and our hearts went into that frequency; problems, hopes, desires, ambitions, doubts, pleasures, fears. KRAB was the crucible of my own ambition; it took me from being a loser poet and failed Washington DC broadcaster to being something of value for my society and my culture. It took me from vague hopes of good programming in 1959 to a purveyor of what is and can be the best in men's souls. It gave me a chance to put out, on the air, a continuing statement in music and words, what people deserve in this fecund society. It gave me a chance to pit my words and my ideas with the thousands that came in the door and window, through mail and telephone. It gave me a chance to sharpen my own critical facilities and learn (it was my college and my graduate school), how to operate a frequency that would give maximum benefit to the maximum numbers of the curious, the hopeful, the wondering. KRAB took me out of a limited orbit of self-doubt, and put me in a place where I could change and delight many—not the least of all, myself. It taught me my power and my glory, and my weaknesses.

Even in that leather and clean-carpet and coffee smelling boardroom, I could remember the stink of KRAB. Too many people in too-small rooms; too many availing themselves of a single toilet, a single control room, a single desk. And about us all—the characteristic whine of the transmitter, the sagging shelves of records from all over the world, the tapes of past words.

Do I miss it? Do I regret the fact that my dream radio station is now being sold for four million smackers? I don't know . . . I just don't know. I am just too far from that past to regret the loss of this or any other station. Too far gone.

The now President of KRAB sits next to me in that board room. His name is Nick Johnson and he doesn't look at all like his namesake: rather he looks like a man who thinks he has spinach on his teeth, and thus refuses to laugh. I decided to make him laugh.

"There's money in that briefcase," he tells me, *sotto voce*.

"Which one?" I say. He is Peter Lorre, I am Clark Gable.

"Either the black one or the brown one," he says.

"The whole four million smackers," I say.

"Yep," he says: "Only there is another briefcase right next to it. And it's got a bomb in it, one that'll blow up the moment you touch it."

"We'll get blown up—and the lawyers too?"

"Yep." I think briefly on the benefits to American society. I advise him to grab the money briefcase and we'll take it to Brazil.

"I'd rather take the ferry to Vashon," says Johnson.

"Shit," I tell him: "With $4,000,000—we can buy the Vashon Ferry and go to Rio . . ."

The attorney for Security Pacific Bank ahems for silence. This is to be the moment, the moment of the passing-of-the-check. There is a religious hush. We are in the *Immaculata* of American business, the time of the handing of the dollars. Bader and Solemn Bank Lawyer stand. I want one of them to say, "This is a moment that will live in perpetuity . . ." but Bank Lawyer merely passes the instrument for $900,000 to Bader. "Could I see your ID?" says Bader to the attorney for the bank which commands $15,000,000,000 in assets.

We laugh: Margason and Johnson and Dawson and I. The lawyer for the opposition essays a weak smile. "Who are these people?" he wonders. Margason, a phlegmatic sort at best, almost breaks his leg getting the $900,000 check to the bank downstairs so he'll have a weekend's worth of interest. This from a man who has been considering (for four years) the cheapest way to install plumbing in his bathroom so he can reconnect the tub. He writes two dozen checks—some to pay off debts from fifteen years ago. He gives me over $27,000 to pay the money Fessenden advanced to the station to keep it going during the wait for closing. "The trouble with the check," I confide to the others, "is that it's no good. Margason misspelled 'thousand.' He's never written a check for that large a sum before."

25 April 1984

MASTER

The light's blown out—where did it go? The darkness is back in the room as before.

ANSWER

The maid is washing clothes at the side of the wall. The servant is pissing in the field.

The Sound of One Hand

THE RADIO PAPERS
WAS KEYBOARDED ON A KAYPRO II MICROCOMPUTER
BY MOLLIE L. FIELD WORKING FROM ORIGINAL PROGRAM GUIDES
AND OTHER SOURCE MATERIALS.

THE TYPESETTING WAS DONE DIRECTLY FROM THE COMPUTER DISCS
BY FRANK'S TYPE
OF MOUNTAIN VIEW, CALIFORNIA
USING A MERGENTHALER PHOTOTYPESETTING MACHINE.

THE TYPEFACE IS KNOWN AS WEISS.
IT WAS NOT DESIGNED BY ANYONE BUT EMERGED, FULLY FORMED,
FROM A SMALL CLOUD OF BLUE SMOKE
WHICH APPEARED ONE BLUSTERY WINTER EVENING
ON THE OUTSKIRTS OF LINCOLN, NEBRASKA.

THE COVER DESIGN AND HAND LETTERING ARE BY JIM PARKINSON.

THIS BOOK WAS DESIGNED BY DOUGLAS CRUICKSHANK.

Lorenzo Wilson Milam